Raphael Poritsky, Ph.D.
Case Western Reserve University
School of Medicine
Department of Anatomy
Cleveland, Ohio

NEUROANATOMICAL PATHWAYS

1984
W. B. SAUNDERS COMPANY
Philadelphia / London / Toronto /
Mexico City / Rio de Janeiro / Sydney / Tokyo

W. B. Saunders Company: West Washington Square
Philadelphia, PA 19105

1 St. Anne's Road
Eastbourne, East Sussex BN21 3UN, England

1 Goldthorne Avenue
Toronto, Ontario M8Z 5T9, Canada

Apartado 26370—Cedro 512
Mexico 4, D.F., Mexico

Rua Coronel Cabrita, 8
Sao Cristovao Caixa Postal 21176
Rio de Janeiro, Brazil

9 Waltham Street
Artarmon, N.S.W. 2064, Australia

Ichibancho, Central Bldg., 22-1 Ichibancho
Chiyoda-Ku, Tokyo 102, Japan

Library of Congress Cataloging in Publication Data

Poritsky, Raphael.

Neuroanatomical pathways.

1. Neuroanatomy—Atlases.   2. Neural circuitry—
Atlases.   I. Title. [DNLM: 1. Neural pathways.
2. Neuroanatomy. WL 102 P836n]

QM451.P6 1984   611.8   83-20452

ISBN 0-7216-1276-8

Neuroanatomical Pathways   ISBN 0-7216-1276-8

© 1984 by W. B. Saunders Company. Copyright under the Uniform Copyright Convention. Simultaneously published in Canada. All rights reserved. This book is protected by copyright. No part of it may be reproduced, stored in a retrieval system, or transmitted in any form or by any means, electronic, mechanical, photocopying, recording, or otherwise, without written permission from the publisher. Made in the United States of America. Press of W. B. Saunders Company. Library of Congress catalog card number 83-20452.

Last digit is the print number: 9   8   7   6   5   4   3   2   1

Dedicated to my wife Connie
and our children, Anne, John and Marc.

ACKNOWLEDGMENTS

I am grateful to the following people who have contributed to this book: Joan Belfer, Debbie Moore, and Carolyn Winn for their efficient and pleasant typing of the manuscript; Ken Kondo for his outstanding photographic work; Dr. Theodore Voneida for his valuable and constructive suggestions; Marion Sherman for her many helpful comments and suggestions; Joseph Kanasz, who drew Figures 24, 45, 46, 47, and 55; Janice Lalikos, who drew Figure 39, and Barbara Davison, who drew Figure 58; Wayne Timmerman for his help in the lettering of the drawings. I am especially indebted to Baxter Venable, Medical Editor of the W. B. Saunders Company, for the editorial assistance and cooperation extended to me in preparing this book; and finally I thank my wife Connie for all her help and support which is deeply appreciated.

I found the following books to be especially useful and the reader should consult these for particular details: Neurological Anatomy in Relation to Clinical Medicine, 3rd ed., Alf Brodal, Oxford University Press, New York, 1981; Gray's Anatomy, 36th British Edition, ed. by Peter L. Williams and Roger Warwick, W. B. Saunders, Philadelphia, 1980; The Human Central Nervous System: A Synopsis and Atlas, R. Nieuwenhuys, J. Voogd, C. van Huijzen, Springer-Verlag, 1978; Structure of the Human Brain: A Photographic Atlas, 2nd ed., Stephen DeArmond, Madeline Fusco, Maynard Dewey, Oxford University Press, New York, 1976; The Human Brain in Sagittal Section, Marcus Singer and Paul Yakovlev, Thomas, Springfield, 1954; Human Brain: An Introduction to Its Function, John Nolte, C. V. Mosby, St. Louis, 1981; The Fine Structure of the Nervous System, Alan Peters, Sanford Palay, Henry deF. Webster, W. B. Saunders, Philadelphia, 1976; Principles of Neurology, Raymond D. Adams and Maurice Victor, McGraw-Hill, New York, 1977; and A Textbook of Neurology, 6th ed., Hiram Merritt, Lea and Febiger, Philadelphia, 1979.

HOW TO USE THIS BOOK

Neuroanatomy is essentially a study of the "wiring" or "circuitry" of the central nervous system. This book is designed to present the essentials of neuroanatomy in a form in which the student can trace out these pathways with a color pencil or color marker. By doing this, the student will discover where the nerve cell bodies lie and where their axons go. This book gives the student the opportunity to play an "active role" and actually trace out and color the pathways. Many students tend to remember things that are learned with a muscular activity more readily than by simply reading about them.

This book is intended primarily for students who wish to learn neuroanatomy. Certain parts of this book and their illustrations may prove useful to instructors, researchers, and students in fields related to neuroanatomy.

The text is placed in the front and the pictures in the back. Each picture may be removed at the time of coloring and inserted into a loose-leaf note book when completed. Two topics, the somatosensory pathways and pyramidal tract, each consists of two figures that must be taped together.

Use either color pencils or color markers. No specific color key is necessary. The student may use any color he or she wishes and the colors used may vary from page to page.

CONTENTS

TITLE	PAGE NUMBER	FIGURE NUMBER
Some Basic Terms	1	Figure 1
More Terms and Concepts	3	Figure 2
The Neuron	4	Figure 3
Nerve and Its Connective Tissue Sheaths	5	Figure 4
Medial View of Brain	7	Figure 5
Inferior View of Brain	7	Figure 6
The Spinal Cord	8	Figure 7
Spinal Nerve Components	10	Figure 8
The Motor Neuron	12	Figure 9
The Motor Unit and Motor Neuron Level	13	Figure 10
The Muscle Spindle	14	Figure 11
Somatosensory Pathways	16	Figure 12 Figure 13
Stereognosis	18	Figure 14
Spinal Cord Lesion	19	Figure 15
The Cranial Nerves	20	Figure 16
Cranial Nerve Components	22	Figure 17
The Hypoglossal Nerve (XII), Accessory Nerve (XI), and Vagus Nerve (X)	24	Figure 18
The Glossopharyngeal Nerve (IX)	26	Figure 19
Central Vestibular Connections (N. VIII)	28	Figure 20
Central Auditory Pathways (N. VIII)	30	Figure 21
Facial Nerve (VII)	32	Figure 22

TITLE	PAGE NUMBER	FIGURE NUMBER
Facial Nerve and Higher Motor Control	33	Figure 23
Bell's Palsy	34	Figure 24
The Trigeminal Nerve (V) and Central Connections	35	Figure 25
The Branchial Arches and the Branchiomeric Muscles	38	Figure 26
The Autonomic Nervous System	40	Figure 27
Basic Pathways of the Autonomic Nervous System	43	Figure 28
Parasympathetic Outflow in the Head	45	Figure 29
Cardiac Pain Fibers	48	Figure 30
Central Visual Pathways (N. V–N. II)	49	Figure 31
Central Connections of the Optic Nerve (N. II)	51	Figure 32
Cranial Nerves VI, IV, and III	53	Figure 33
The Cerebellum—1. General Plan	55	Figure 34
The Cerebellum—2. Cortical Neurons	57	Figure 35
The Cerebellum—3. Afferents, Dorsal View	59	Figure 36
The Cerebellum—4. Afferents, Lateral View	61	Figure 37
The Cerebellum—5. General Circuitry	62	Figure 38
Symptoms of Cerebellar Disease	64	Figure 39
The Pyramidal Tract	65	Figure 40 Figure 41
The Caudate Nucleus, Putamen, Globus Pallidus, and Amygdala	68	Figure 42
The Basal Nuclei. Interconnections 1	70	Figure 43
The Basal Nuclei. Interconnections 2	71	Figure 44
Parkinson's Disease	72	Figure 45
Chorea	73	Figure 46
Athetosis and Hemiballism	75	Figure 47
Stria Terminalis and Ventral Amygdalofugal Pathway	76	Figure 48
Some Afferent Connections in the Amygdaloid Complex	78	Figure 49
The Hippocampal Formation	79	Figure 50

TITLE	PAGE NUMBER	FIGURE NUMBER
The Hippocampus and Fornix	80	Figure 51
The Internal Organization of the Hippocampal Formation	81	Figure 52
Olfactory Pathways	82	Figure 53
The Thalamus—1. Anterior and Medial Nuclei	84	Figure 54
The Thalamus—2. Lateral and Posterior Nuclei	85	Figure 55
The Cerebral Cortex	86	Figure 56
The Strange Case of Phineas P. Gage	92	Figure 57
Nursing Pathways		Figure 58

Some Basic Terms   Figure 1

A. Upper Figure

1a. <u>Nucleus</u>* - A <u>nucleus</u> in neuroanatomy is a group of nerve cell bodies inside the brain or spinal cord. The brain and spinal cord comprise the CNS or central nervous system. Hence, a <u>nucleus</u> may also be defined as "an aggregation of neuronal cell bodies inside the CNS." A nucleus may be so small that it consists of a few hundred neurons; or it may consist of millions of neurons and be so large that it is readily seen with the naked eye. Six neuronal cell bodies (without their dendrites) are shown in the nucleus at "1a". The word "nucleus" is derived from the Latin <u>nux</u>, (nut, kernel) and originally meant "little kernel".

1b. <u>Efferent fibers</u> - <u>Efferent</u> means outgoing or traveling away from the point of reference. The fibers "1b" (or axons) arising from the cell bodies at "1a" carry nerve impulses that travel away from the CNS. The neuronal cell bodies at "1a" are the <u>cells of origin</u> or parent cell bodies of the fibers at "1b". <u>Motor</u> means causing movement. Fibers such as those in "1a" would be motor if they conveyed impulses that resulted in the contraction of muscle.

2a. <u>Ganglion</u> - A <u>ganglion</u> is a collection of nerve cell bodies outside the CNS. Notice that these cell bodies are connected to their fibers by a single process. Hence, the neuronal cell bodies at "2a" are <u>unipolar</u> which means they have only one process. Notice also that the fibers travel through this ganglion with no break or synapse.** All nerve fibers that convey sensations such as pain, touch and temperature plus many that are concerned with autonomic reflexes at the unconscious level have parent cell bodies similar to those shown at "2a".

2b. <u>Afferent fibers</u> - <u>Afferent</u> means incoming or traveling towards. In neuroanatomy, connections are usually described in terms of afferent and efferent fibers; that is, what fibers bring impulses to a certain structure and what fibers carry impulses away from this structure. <u>Sensory</u> means conveying sensation. Actually many "sensory fibers" carry information that will not reach the conscious level, and, hence, are better termed "afferent fibers".

3. <u>Mixed nerve</u> - A <u>nerve</u> is a collection of neuronal fibers or axons outside the CNS. A mixed nerve such as shown in "3" contains both motor and sensory fibers.

4a. <u>Second order</u> - At "4a" the afferent fibers "2b" end and synapse upon a group of neurons (or nucleus) that comprise the second order neurons. This simply means they are the second link or leg in series of neurons that together comprise a whole pathway within the CNS. The neurons at "4a" give rise to <u>second order</u> fibers at "4b" that project to higher centers in the CNS.

5. <u>Tract or fasciculus</u> - This is a collection of nerve fibers or axons inside the CNS. The earlier anatomists often gave these structures fanciful names such a <u>lemniscus</u> (Latin, a fillet) for a ribbon-like tract, or <u>ansa</u> (Latin, handle) for a curved tract. Additional names include <u>peduncle</u> (Latin, little foot), <u>corpus</u> (Latin, body) and <u>brachium</u> (Latin,

arm). The important thing to remember is that these are all names for bundles of axons inside the CNS. (Corpus is also used for some nuclei.)

B. Lower Figure

A "disassembled" neuron showing its component parts.

6. <u>Soma or perikaryon</u> - Both these terms mean neuronal cell body. <u>Soma</u> (plural, somata) is Greek for <u>body</u>. <u>Peri</u> (Greek, around) plus <u>karyon</u> (Greek, kernel) form <u>perikaryon</u> which literally means "that which is around the kernel". This neuron is <u>multipolar</u> since it gives rise to more than two processes (that is an axon and at least two dendrites).

7. <u>Dendrite</u> - Dendrites are processes of the neuron that arise from the soma and greatly increase the receptive surface of the neuron. They tend to branch much like tree limbs; hence, they were named <u>dendrites</u> which is derived from the Greek <u>dendron</u> which means tree or branch. Only the proximal parts of four dendrites are shown here.

8. <u>Axon hillock</u> - This is a cone-shaped projection of the perikaryon from which the axon arises.

9. <u>Initial segment</u> - This is the first part of the axon. It has no myelin sheath. The nerve impulse begins at this part of the axon.

10. <u>Axon</u> - The <u>axon</u> or <u>nerve fiber</u> or simply <u>fiber</u> carries the nerve impulse. This particular fiber is a myelinated fiber since it has a covering of myelin.

11. <u>Myelin</u> - The myelin sheath forms a layer of insulation that greatly increases the axon's conductive velocity.

12. <u>Node of Ranvier</u> - Interruptions in the myelin sheath are called nodes of Ranvier. At "12" the axon divides or <u>bifurcates</u> and gives off a side branch or <u>collateral</u> (14).

13. <u>Myelinated axon</u> - Also called medullated axon. A fiber ensheathed with myelin.

14. <u>Collateral</u> - This is a side branch of the axon.

15. <u>Recurrent collateral</u> - This is a side branch that runs back toward the cell of origin.

16. <u>Axon terminals</u> - They are small swellings at the ends of axons. They are also called synaptic bulbs, terminal buttons, and boutons terminaux (French, terminal buttons).

<u>Directions</u>: With a sharp color pencil, color the <u>efferent</u> fibers (1b) and their cells of origin (1a) the same color as the square marked <u>efferent</u>. Do the same for the <u>afferent</u> fibers (2b) and their cells of origin (2b) and the square <u>afferent</u> using another color.

---

*These figures are not drawn to a single scale.

**Ganglia of the autonomic nervous system do, however, contain synapses.

More Terms and Concepts   Figure 2

A. Descending Tracts

   1.   Homolateral - The somata indicated at "1" give rise to fibers that remain homolateral. That is, they remain on the same side as their cells of origin. Ipsi lateral means the same as homolateral.

   2.   Contralateral - The fibers that arise from the perikarya labelled "2" cross the median plane at point "3" and become contralateral or crossed fibers.

   3.   Decussation - Decussate means to cross to the opposite side. The nervous system, like the whole body, is bilaterally symmetrical. Wherever one tract crosses, its counterpart will also cross. Tracts tend to cross and intersect each other in an oblique direction thus forming an "X". Decussis in Latin meant: to cut in the form of an "X".

   4.   Lower motor neurons - These somata reside in either the brain stem or spinal cord. Their axons extend into nerves and eventually end upon striated muscle. In this picture they receive synapses from the descending fibers originating from somata labelled "1" and "2".

B. Ascending Tracts

   1.   First order sensory neuron - This term means simply that this is the first or number-one neuron in a chain of neurons that carry a particular sensory modality (or kind of sensation). Such a chain of neurons is also referred to as a pathway.

   2.   Second order neuron - This neuron synapses with two third order neurons.

   3.   Third order neuron - One of these neurons sends its axon to the opposite side where it forms a contralateral ascending tract. The other third order neuron's axon remains homolateral.

   4.   Fourth order neurons - These receive synapses from third order neurons and project their axons to fifth order neurons in the cortex. (Actually the identification of orders greater than one is not always possible due to the existence of small interneurons.) The bottom figures (C, D) are typical "wiring" diagrams showing the same pathways as the figure above each. Each cell and its axon might represent hundreds, thousands, or even millions of neurons. Note how the synapse is indicated.

Etymology is the study of word origins. An appreciation of the origin of biological and anatomical terms usually helps in understanding and remembering the term.

Directions: Color the neurons in figures A and B as indicated in figure 2.

The Neuron   Figure 3

1. <u>Axodendritic synapses</u> - Each of these show a small portion of an afferent fiber and its axon terminal (or terminal button or bouton terminau) forming a synapse up a dendrite.

2. <u>Axosomatic synapses</u> - Two such synapses are shown here.

3. <u>Axoaxonic synapse</u> - Shown here is the terminal of one axon ending upon the beginning of another axon thereby forming an <u>axoaxonic</u> synapse.

4. <u>Synapse in passing</u> - This type of synapse is formed by an axon contacting and synapsing upon the target cell at some length before it ends. Synapses formed in this manner are called synapses in passing (or <u>en passant</u>).

5. <u>Nissl substance</u> - This is actually the granular endoplasmic reticulum in nerve cells. It may be block-like, as shown here, or take other forms. It consists of cisterns and ribosomes. That latter contained RNA and account for the basophilia of the Nissl substance.

6. <u>Nucleus</u> with nucleolus.

7. <u>Oligodendrocyte</u> - One of the neuroglial cells.

8. <u>Astrocyte</u> - Another type of neuroglia.

9. <u>Astrocytic processes</u> - These extend from the astrocyte cell body to the neuron. Astrocytes have exceedingly convoluted surfaces.

10. <u>Perivascular end feet</u> - These are formed by astrocytic processes that end upon capillaries in the CNS.

11. <u>Capillary</u> - Capillaries in the CNS are almost completely ensheathed by perivascular end feet from astrocytes.

12. <u>Axon</u> - The cytoplasm within the axon is called <u>axoplasm</u>. The cell membrane about the axon is the <u>axolemma</u>.

13. <u>Oligodendrocytic cytoplasm</u> - Myelin within the CNS is made by the oligodendrocyte. A thin "tongue" of oligodendrocytic cytoplasm resides both within the myelin sheath and on its external surface.

14. <u>Myelin sheath</u> - Spirally-wrapped oligodendrocyte plasma membrane. In nerves, that is, outside the CNS, it is formed by spirally wrapped Schwann cell membrane.

Nerve and Its Connective Tissue Sheaths   Figure 4

Figure A shows a nerve consisting of several fascicles (4). Figures B, C, and D are successive enlargements of one fascicle (B), one axon (C), and part of an axon (D).

1. <u>Epineurium</u> - This is the outermost connective tissue sheath about a nerve. It is mainly collagenous fibers.

2. <u>Perineurium</u> - This is the connective tissue about each fascicle. Previously thought to be largely collagenous fibers, now considered to consist of several layers of cells, the "perineurial epithelium," alternating with layers of collagenous fibers and probably forming a selective barrier about each fascicle. Probably continuous to the end of each fiber.

3. <u>Endoneurium</u> - Loose aggregation of collagenous fibers about each nerve fiber. Believed to play an important role in guiding regenerating axons.

4. <u>Fascicles</u> - Large nerves consist of several fascicles or bundles of axons.

5. <u>Axon</u> - Its cytoplasm is <u>axoplasm</u>. Its cell membrane is the <u>axolemma</u>.

6. <u>Myelin</u> - In nerves myelin is made by the spiraling and condensing of the Schwann cell membrane.

7. <u>Schwann cell</u> - The Schwann cell forms a cellular sheath external to the myelin. This sheath is also called the neurilemma (or sheath of Schwann). One Schwann cell ensheaths an axon between two nodes of Ranvier (or for each <u>internodal</u> segment). A thin layer of Schwann cytoplasm (7') is found squeezed between the axon and the surrounding myelin. Schwann cells are not present in the CNS. Oligodendrocytes make myelin in the CNS.

8. Capillaries.

9. Arteries and veins.

10. <u>Basal lamina</u> - A thin noncellular layer of ground substance.

11. Membrane-enclosed vesicles believed to be derived from the smooth endoplasmic reticulum within the perikaryon.

12. <u>Microtubules</u> - These are found in the axon, dendrites, and perikarya. They are about 240 angstroms in diameter (24 nm).

13. <u>Neurofilament</u> - Thinner than microtubules, about 80 angstroms (8 nm), these are also found in the axon, dendrites and perikaryon.

14. <u>Mitochondrion</u> - These and other organelles actually move down the axon (in a proximal-to-distal direction) apparently the result of an "axoplasmic flow."

---

*Remember, by definition a "nerve" is never within the brain or spinal cord; rather it is a collection of nerve fibers <u>outside</u> the CNS. In addition to axons and myelin sheaths, larger nerves contain blood vessels, connective tissue sheaths and cells such as fibroblasts and macrophages.

Microscopic units of length are:

    the millimeter (mm); (1000 mm = 1 meter)

    the micrometer (μm); (μ or micron is the old name); 1000 μm = 1 mm

    the nanometer (nm); 1000 nm = 1 μm

    the angstrom (A); 10 A = 1 nm, or 10,000 A = 1 μm

Medial View of Brain Figure 5

Directions: Color each of the structures indicated by the asterisks (*).

Inferior View of Brain Figure 6

Directions: Color each of the structures indicated by the asterisks (*).

The Spinal Cord   Figure 7

A. Upper figure

The spinal cord and brain are covered by three connective tissue membranes called meninges (Greek, membrane). The innermost meninx is the pia mater (Latin, pia means either gentle, tender or pious; mater is mother). It is the thinnest and faithfully covers the surface of the brain and spinal cord extending the full depth of each fissure and sulcus (Latin, groove).

The pia contains blood vessels and laterally forms two denticulate ligaments on each side of the spinal cord that help stabilize the spinal cord within the vertebral canal. The middle meninx is the arachnoid mater (Greek, spider or web). It is so named because of the many tiny fibers or arachnoid trabeculae (Latin, little beams) that extend across the subarachnoid space from the pia to the arachnoid. The considerable subarachnoid space contains cerebrospinal fluid which protects the brain and spinal cord with its aqueous environment. Enlargements of the subarachnoid space are called cisterns (cisternae). The cerebrospinal fluid supports the brain and spinal cord against gravity and more evenly distributes blows and concussions to the central nervous system.

The outermost meninx is the dura mater (Latin, tough mother). It is fibrous and strong and extends laterally to ensheath each dorsal root ganglion. The dura mater becomes continuous with the epineurium of the nerve at the lateral border of the ganglion.

The inner gray matter which contains largely neuronal somata in cross section forms the "H" or butterfly with a larger ventral horn (v horn) and a somewhat thinner dorsal horn (d horn). There are 31 pairs of spinal nerves. Each spinal nerve arises by a ventral root and a dorsal root. (In the human anterior and posterior are synonymous with ventral and dorsal, respectively.) Note that each root is formed by the coalescence of many dorsal and ventral root filaments.

Each spinal nerve divides into a large ventral and smaller dorsal ramus (Latin, branch). The ventral rami (also called ventral primary ramus) become the "nerves" of the body such as the ulnar, median, femoral, and intercostals. The dorsal rami innervate only the deep muscles of the back such as the erector spinae and the skin of the back.

In the brachial and lumbosacral plexuses, the nerve fibers become divided into either ventral and dorsal divisions. In the brachial plexus, the nerves then regroup to form three cords and finally the large nerves of the arm issue from these cords. In primitive vertebrates, the ventral division supplied the ventral flexor (flex) muscles of the limbs and the dorsal division supplied the dorsal extensor (ext) muscles. This still holds true in the arm where the radial nerve, which is derived from the dorsal division and dorsal cord, supplies all the extensor muscles of the arm. The rotation of the limbs and the upright stance have somewhat distorted this pattern in the leg; for instance, a muscle such as the iliopsoas which flexes the thigh at the hip originally was an extensor. The original role of this muscle is revealed by its nerve, the femoral, which is derived from the dorsal division of the lumbar plexus. The tibial nerve, on the other hand, is derived from the ventral division and supplies most of the flexors of the leg.

B. **Middle figure**

Note that the ventral ramus and dorsal ramus each contain both motor and sensory fibers. Motor neuron "1" travels with the ventral ramus and supplies a muscle such as an intercostal or a limb muscle. Motor neuron "2" supplies a deep muscle of the back such as the erector spinae via a dorsal ramus. In the fish where this division of muscles is more nearly 50-50, the dorsal trunk musculature is the epaxial and the ventral musculature is the hypaxial. All fin (and limb) musculature is supplied by ventral rami. Note that sensory neurons "3" and "4" with their unipolar perikarya in the dorsal root ganglion supply sensory fibers to both the ventral ramus (fiber "3") and dorsal ramus ("4"). Fiber "4" enters the spinal cord via the dorsal root and synapses in the dorsal horn upon internuncial neuron "5" which synapses upon motor neuron "2". Afferent fiber "3" terminates upon neuron "6" which is a cell of origin of a crossed ascending tract. Fiber "7" is a descending fiber from a higher center (that is, the brain). It ends upon interneuron "8" which in turn synapses upon motor neuron "9". Remember that it is only roots which contain either motor or sensory fibers and that the rami are mixed and contain both motor and sensory fibers.

C. **Lower figure**

The white matter consists mainly of longitudinally oriented myelinated axons. In the spinal cord, the white matter is divided into three funiculi (Latin, little rope), the ventral, lateral, and dorsal funiculi, shown on the left side. On the right side, it is apparent that each funiculus contains several tracts or fasciculi (Latin, little bundles). The dorsal funiculus contains two tracts or fasciculi, the fasciculus gracilis and fasciculus cuneatus. The layers or laminae of Rexed are shown on the left side. Rexed, a Swedish neuroanatomist, on the basis of cell types, divided the gray matter into ten laminae. Layer I is most dorsal. Layer II is the substantia gelatinosa (Latin, jelly-like substance) (sg on right). Note that lamina IX which contain the motor neurons is not a "lamina" at all, but rather forms several columns. Lamina X contains the cells around the central canal. The ascending and descending tracts in the white matter cannot be identified except by special means.

Directions:

Upper figure. Color each of the meninges a different color. Color the denticulate ligament the same color as the pia, and the trabeculae the same as the arachnoid. Color the gray matter another color and use a different color for the ventral and dorsal root filaments.

Middle figure. Use one color for motor neurons "1" and "2" and another for sensory neurons "3" and "4". Color the white and gray matter each and trace the connections involving neurons "5" through "9".

Bottom figure. Color each of the three funiculi on the left. Then color lamina I, III, V, VII, and IX.

Abbreviations: d horn, dorsal horn; dorsolat fasc, dorsolateral fasciculus; dorsal fun, dorsal funiculus; dorsal sp cer tr, dorsal spinocerebellar tract; fasc cun, fasciculus cuneatus; fasc gr, fasciculus gracilis; lat fun, lateral funiculus; lat cort sp tr, lateral corticospinal tract; propriosp tr, propriospinal tract; rc, rami communicantes; sp th trs, spinothalamic tracts; sg, substantia gelatinosa; vent fun, ventral funiculus; v horn, ventral horn, vent sp cer tr, ventral spinocerebellar tract.

Spinal Nerve Components   Figure 8

GSE - **General somatic efferent.** This includes the axons of alpha motor neurons and axons of gamma motor neurons.

GVE - **General visceral efferent.** This component includes all sympathetic and parasympathetic fibers, both preganglionic and postganglionic.

GSA - **General somatic afferent.** This component includes all fibers from exteroceptors (that is, surface receptors such as pain, touch, and thermal receptors) plus fibers from all proprioceptors (that is, position and movement sensors).

GVA - **General visceral afferent.** This includes pain fibers, hunger fibers, and afferent fibers that play vital roles in reflexes, most of which operate below the conscious level.

1. **Alpha motor neuron.** Also called lower motor neuron. Its axon innervates voluntary skeletal muscles. (See figure 9.)

2. **Gamma motor neuron.** Its axon terminates upon the intrafusal cells in muscle spindles. (See figure 11.)

3. **Preganglionic sympathetic neuron.** It lies in the intermediolateral column of the spinal cord. Its axon ends upon postganglionic neuron "7".

4. **Muscle spindle cell.** The gamma motor neuron's discharge causes the ends of this intrafusal fiber to contract. (See figure 11.)

5. **Striated muscle cell.** Also referred to as extrafusal fiber in contradistinction to the intrafusal fiber that composes the muscle spindle.

6. **Gland.** Besides glands, the heart (cardiac muscle) and smooth muscle are also innervated by the terminals of postganglionic autonomic fibers.

7. **Sympathetic postganglionic soma** residing in a ganglion of the sympathetic chain. Its axon (7') runs into the spinal nerve where it will eventually end upon either a hair follicle, sweat gland, or blood vessel. (See figure 27.)

8. **Sympathetic ganglion** contains postganglionic sympathetic perikarya.

9. **Dorsal root ganglion** contains the unipolar perikarya of first order afferent neurons.

10. **Soma of GVA fiber.** General visceral afferent fibers carry sensation such as pain, nausea, and hunger. These sensations tend to be diffuse and not specifically localized.

11. **Soma of GSA fiber.** This is the perikaryon of a fiber that carries either pain, touch, temperature, or proprioception.

12. **Meissner's touch corpuscle.** This is a tactile (or touch) receptor located in the dermal papilla. It responds to any slight deformation of the overlying epidermis. (See figure 14.)

13. Intestine (with villi).

<u>Directions</u>: Color each of the four spinal nerve components.

The Motor Neuron Figure 9

The motor neuron, or more correctly, the alpha motor neuron is located in the ventral horn of the spinal gray matter. It is also found in the motor nuclei of certain cranial nerves. In addition to being called "alpha motor neuron," it is also known as the "lower motor neuron," or "peripheral motor neuron," and Sherrington called it "the final common pathway."

It is a very large cell with the diameter reaching 100 m or more. It has several large dendrites which often extend for considerable distances (as much as 1000 μm). Its perikaryon and dendrites are covered with the axon terminals (boutons terminaux) of afferent fibers. It is estimated that a single motor neuron may receive as many as 30,000 boutons on its perikaryon and dendrites. The drawing shows only a few boutons of afferent fibers. Actually the surface is so covered with boutons that if all the incoming fibers were cut and removed and only the boutons were left on the surface, it would resemble a clove-studded pomander ball.

Note the nearby astrocyte (Greek, star cell) which is one of the neuroglial cells. Note how irregular and convoluted its surface is due to its many processes squeezing between and about all the other structures in the central nervous system. The astrocytic processes often squeeze between boutons in such a way that they appear to prevent neighboring boutons from exciting each other and thus act as insulators. Astrocytic processes also extend to capillaries in the CNS where they form perivascular end feet which almost completely ensheath these small blood vessels. This strongly suggests some sort of selective barrier role and/or transport function for the astrocyte, but this has not been demonstrated.

Note the two oligodendrocytes (Greek, few branch cell). The perineuronal oligodendrocyte is located next to the motor neuron where it appears to metabolically assist the motor neuron. The other oligodendrocyte is shown with five processes that extend to axons. The oligodendrocyte synthesizes myelin in the CNS. Unlike its counterpart, the Schwann cell which synthesizes myelin in the peripheral nervous system for only one internode on a single axon, one oligodendrocyte may form the myelin for as many as 40 internodes of nearby axons.

The neuroglial cells, which consist of astrocytes, oligodendrocytes, and a third type of cell, the microgliocytes, greatly outnumber the neurons with some estimates as high as 10 neuroglial cells to each neuron.

Directions: Color the motor neuron, the astrocyte, and the oligodendrocytes each a different color.

The Motor Unit and Motor Neuron Level   Figure 10

The motor unit consists of an alpha motor neuron, its axon, and all the muscle fibers innervated by that motor neuron. Each motor unit works as a functional unit. Figure A shows an alpha motor neuron innervating 8 muscle fibers. This is a very small number of muscle fibers and is typical of small muscles that are capable of precise and delicate movement such as the muscles of the fingers and extrinsic eye muscles. However, large muscles such as the gastrocnemius have motor units in which one motor neuron may innervate up to 1,800 muscle fibers. Muscles with such large motor units as the gastrocnemius have relatively fewer motor units and are not able to contract with as many different amounts of tension as are muscles with more numerous small motor units. The force of the contraction depends upon how many motor units are simultaneously discharging and the frequency of impulses from the motor neuron.

Directions: Color the motor unit (motor neuron and the muscle fibers innervated).

Figure B shows the motor neurons that supply shoulder muscles lie higher in the spinal cord than do motor neurons that supply the more distal limb musculature. Thus the motor neurons that innervate the levator scapulae lie at levels C_3-C_4; whereas those to the biceps lie at C_5-C_6; and those to the adductor pollicis lie at C_8-T_1. The importance of this is evident when the spinal cord is damaged in the neck region. If the break is just above C_8, the individual will lose hand function, but still have use of the more proximal parts of the arm. Or if the spinal cord were severed just below C_4, only the most proximal limb muscles such as the levator scapulae would be able to function. The loss of function is due to the interruption of descending motor fibers in the spinal cord white matter.

Directions: Color each of the motor neurons, their fibers, and the muscles receiving innervation.

The Muscle Spindle   Figure 11

The muscle spindle is a length sensor. It signals information on the length of the muscle, on any change in length, and on how fast the change occurs. Figure A is a single cell (or fiber) from a muscle spindle.

Each muscle spindle in man contains an average of 10 such cells called <u>intrafusal</u> cells. The rest of the muscle fibers, that is, those that actually contract, are called <u>extrafusal</u> cells (or fibers). Each muscle has many spindles. The small abductor pollicis brevis has about 80 and the large latissimus dorsi over 350.

In figure A the spindle cell is in a stretched state and in figure B the same cell is in a shortened state. The stretching or lengthening of a spindle occurs when its muscle, such as the biceps brachii (figure C) is lengthened. Notice that the spindle cell in figures A and B has the annulospiral ending of a la afferent fiber wrapped around its central sensory portion. The annulospiral ending is very sensitive to any stretching of the central portion of the spindle cell. This central part contains a number of nuclei (not shown) and is non-contractile; that is, it cannot contract by itself. The contractile portions of the spindle cell are at its ends, and receive their motor supply from gamma ( ) motor neurons.

Discharge of the gamma fiber will cause the contractile ends of the spindle cell to contract which will stretch the central sensory portion with its annulospiral ending. Thus, the gamma motor neuron determines the sensitivity of the spindle by maintaining its sensory portion at the critical threshold. However, the spindle fires off not only when the muscle is lengthened, but also discharges at a steady low frequency even when the muscle is "at rest," that is, not changing its length.

<u>Color</u> the annulospiral ending about the spindle cell in figure A and its afferent fiber. Its unipolar perikaryon is in the spinal (or dorsal root) ganglion and its central portion synapses directly upon alpha motor neurons of the same muscle (seen in figure C "3").

<u>Color</u> the gamma motor neuron in both figures A and B. Its perikaryon is in the ventral horn of the spinal cord gray matter in the vicinity of alpha motor neurons. Its axon leaves the spinal cord in the ventral root and travels to the target muscle in the same nerve as the alpha motor axons.

Appreciate the fact that the gamma motor neuron cannot either lengthen or shorten the spindle; that is determined by the length of the muscle itself. The gamma motor fiber "tightens up" the central sensory part of the cell. The spindle is <u>not</u> subjected to the tension within the muscle. It is arranged <u>in parallel</u> with the extrafusal fibers.

Figure C shows the biceps and its spindle lengthened as occurs with extension of the elbow.

<u>Color</u> the alpha motor neuron and its fiber to the triceps.

Figures C and D show a large isolated spindle cell below the biceps and the same cell within the biceps, both of which are attached to the same la afferent. Starting with the annulospiral endings around both spindle cells, <u>color</u> the la fiber centrally to the spinal cord where it divides into a number of branches. Three of the terminal branches of the la afferents are shown in figure C. These include (1) the dorsal columns (fasciculus cuneatus), (2) an inhibitory interneuron that inhibits the triceps alpha motor neuron, and (3) a synapse directly upon the alpha motor neuron to the same muscle (the biceps brachii).

The synapse of Ia spindle afferents upon biceps alpha motor neuron(s) constitutes a two-neuron monosynaptic reflex and is the basis for the myotatic or stretch reflex (such as tapping the patellar tendon).

Figure D shows the biceps contracted and its muscle spindle shortened. This occurs when its alpha motor neurons (agonists) discharge. In normal movement the contraction of any single muscle probably requires tens or hundreds of motor neurons to discharge, each causing the contraction of all the muscle fibers that it innervates (i.e., its motor unit).

When a muscle contracts, not only do the alpha motor neurons discharge but the gamma motor neurons discharge simultaneously. This coactivation of gamma and alpha motor neurons insures a continuous sensitivity for the spindle so that it does not become "unloaded" or "desensitized" whenever it is shortened. This coactivation is also a means of increasing the alpha motor neuron's rate of firing and, hence, the force of contraction.

In figure D, color both the alpha and gamma motor neurons and their fibers. The gamma fiber is causing the ends of the spindle cell to contract.

Somatosensory Pathways   Figures 12 and 13

Directions: Cut out figures 12 and 13, carefully align them, and tape them together so that the tracts continue uninterrupted from one page to another. Starting with "1pp" in the lower left, color the course of this afferent fiber which carries either conscious proprioception, discriminative touch, or vibratory sense from receptors in the leg such as joint receptors, Meissner's corpuscles, and Pacinian corpuscles. "1pp" is the peripheral process of this first order afferent neuron and could easily extend over a meter in length if it runs from, for example, a joint receptor in the big toe to the dorsal root ganglion of spinal nerve L_5. "1p" is the unipolar perikaryon of this neuron lying in the dorsal root ganglion. The central process (1cp) turns medial upon entering the spinal cord and ascends in the dorsal columns forming a tract or fasciculus, the fasciculus gracilis (1fg). Trace and color the ascent of fiber "1". Note that it remains dorsal, medial, and homolateral (not crossed). It ends in the nucleus gracilis where it synapses upon second order neurons.

Fiber "2" carries similar sensory modalities from the arm. The central process of neuron "2" also turns medial upon entering the spinal cord and ascends as the more lateral fas- ciculus cuneatus (2fc). Follow fiber "2" upwards until it too terminates in another nucleus in the medulla, the nucleus cuneatus. Thus the dorsal columns (or dorsal funiculus) are composed of the central processes of first order sensory neurons whose perikarya reside in dorsal root ganglia.

Locate neurons "3o" and "4o" in the nuclei gracilis and cuneatus. These are the second order neurons and are the cells of origin of the internal arcuate fibers (3iaf, 4iaf) and the medial lemniscus (Med lemn).

Color these fibers and note that the internal arcuate fibers decussate in the medulla and form a ribbon-shaped tract, the medial lemniscus that ascends to the thalamus. Thus the cells of origin (3o, 4o) of the medial lemniscus reside in the nuclei gracilis and cuneatus and give rise to fibers that cross the midline and become contralateral (3, 4). Trace the medial lemniscus superiorly until it reaches the ventralis posterolateralis nucleus (nuc vpl) of the thalamus where its fibers end and synapse upon third order neurons in this nucleus. The somata in the vpl nucleus (5o, 6o) give rise to fibers that project to the postcentral gyrus of the cerebral cortex. This strip of cortex which is immediately behind the central sulcus (or fissure of Rolando) is also called the somatosensory cortex, areas 3, 1, 2 of Brodmann and SmI (primary Sensorimotor cortex). Note that the leg fiber ends superior and medial; whereas the arm fiber projects more laterally.

Locate sensory fiber "7" in the lower right. This fiber carries either pain or thermal sense. Its perikaryon is also unipolar and located in the dorsal root ganglion. Pain and temperature fibers terminate in a number of sites in the gray matter. The cells of origin of the spinothalamic tracts shown in the diagram as "8o" and "14o" have not been positively identified but evidence suggests that they probably lie more ventrally than shown here. In the case of pain transmission, it is believed that one or more neurons are interposed between the endings of the first order neurons and the cells of origin of the lateral spinothalamic tract.

Follow the course of fiber "8". Note that it decussates in the anterior white commissure, usually within one segment and ascends as the lateral spinothalamic tract. Locate fibers "10" and "11" bringing in pain and temperature sense from the arm. Note that pain fibers from the arm (11) assume a more medial (or internal) position in the lateral spinothalamic tract. This layering of sensory fibers from different parts of the body is called a "somatotopic lamination" and is characteristic of the major sensory pathways.

Unlike the dorsal-columns-medial-lemniscus pathway, which only consists of three neurons linked together, the spinothalamic pathways contain many different types of neurons not readily identifiable as first, second, or third order neurons, plus many synapses and interneurons located along these pathways.

<u>Trace</u> pain fibers "8" and "11" up the lateral spinothalamic tract until they reach the vpl nucleus of the thalamus. They also terminate in this nucleus, although in a more diffuse manner than those of the medial lemniscus. Neurons in the vpl nucleus (9o, 12o) relay this information to the somatosensory cortex.

The disagreeable affective quality of pain is probably felt and experienced at the thalamic level. The cerebral cortex appears to be essential in the cognitive evaluation of the nature of injury, such as its extent and precise location.

Return to the lower right and locate fiber "13" carrying "light touch" (tested by gently stroking the skin with a wisp of cotton). First order fibers carrying light touch synapse upon neurons (14o) in the intermediate gray matter which in turn give rise to axons that cross the midline and form the <u>ventral spinothalamic tract</u> (14).

The two spinothalamic tracts, the lateral and ventral, eventually join together. They are sometimes referred to as a single entity, the "anterolateral funiculus". Somata (15o) in the vpl nucleus then relay "light touch" information to the somatosensory cortex (15).

Afferent fiber "16" carries discriminative touch from encapsulated endings in the face. It synapses upon second order neurons (17o) in the principal (or main) sensory nucleus in the pons. These neurons have axons (17) that decussate and assume a position somewhat dorsal to the medial lemniscus. They ascend and terminate in the nucleus ventralis posteromedialis (18o) (nuc vpm) of the thalamus which in turn relays this information to the "face" region of the somatosensory cortex (18). Neuron "19" is a motor neuron in the motor nucleus of V. Its axon exists alongside the larger sensory portion of the root of nerve V. These motor neurons and axons supply the muscles of mastication and comprise the portio minor of the root of nerve V, whereas the more abundant sensory fibers comprise the much larger portio major. Neuron "20" carries either pain or temperature and turns caudal upon entering the pons. These fibers group together into a bundle, the spinal tract of V, and terminate with synapses upon neurons in the spinal nucleus of V (21). (See figure 25.)

Abbreviations: fasc cun, fasciculus cuneatus; fasc gr, fasciculus gracilis; lat sp th tr, lateral spinothalamic tract; med lemn, medial lemniscus; motor nuc V, motor nucleus of nerve V; nuc vpl, nucleus ventralis posterolateralis; nuc vpm, nucleus ventralis posteromedialis; pr sen nuc V, principal sensory nucleus of V; semil gang, semilunar ganglion; sp nuc V, spinal nucleus of V; sp tr V, spinal tract of V; vent sp th tr, ventral spinothalamic tract.

---

*Pathways are only shown on one side for reasons of space.

Stereognosis Figure 14

Stereognosis (Greek, literally "solid knowledge", from _stereos_, solid, and _gnosis_, knowledge) is the ability to identify objects by touch. This ability depends upon several kinds of sensory receptors, such as the Meissner's touch receptor shown in figure 14, plus the brain's extraordinary capacity to combine information from various receptors, each with their own pathway to the cortex and each supplying its own modality of sensation, whether it be fine touch, joint movement, or pressure. Presumably the parietal cortex, in close cooperation with certain thalamic nuclei, processes incoming tactile and proprioceptive information and "internally reconstructs" the object being examined by one's fingers so that the individual can readily identify what he is touching.

Since this "solid sense" relies upon muscle activity and movement of the fingers, a more appropriate name for this kind of touch should be "active touch" rather than "fine" or "discriminative" touch which are terms traditionally used.

Meissner's tactile corpuscles are particularly abundant in the tips of the fingers where one in every four dermal papillae contains one (figure B). In figure C the zig-zig course of the receptor axon is shown as it winds its way between the shelf-like connective tissue cells that form the bulk of the corpuscle.

Several nerve fibers are found in each corpuscle. Whether these are actually separate fibers or whether they are all branches of the same parent fiber is not known. Note the collagen fibers that connect the corpuscle to the overlying epidermis. Because of these collagen fibers, even the slightest movement in the epidermis, such as might occur in feeling a coin, would cause a displacement of the corpuscle and a discharge of its receptor axon.

By itself, the Meissner's corpuscle could not inform one as to the size or dimension of the object being held. This would depend on other receptors, particularly the joint receptor which provides information on just how far the fingers are flexed and extended as the finger tips probe the object. Knowledge of the extent of the finger movement is essential in gauging the diameter of a coin.

Stereognosis is believed to ascend in the cuneate fasciculus and medial lemniscus to the thalamus where it is processed and then relayed to the parietal cerebral cortex where its identity and nature are recognized. Damage to any part of this pathway such as the cuneate fasciculus, medial lemniscus, or parietal cortex will result in _astereognosis_; that is, the inability to recognize objects by touch.

Directions: Color the receptor axon in figure C and the Meissner's corpuscle in figure B.

Figure C is based upon and modified from Andres and von During, Morphology of Cutaneous Receptors in Handbook of Sensory Physiology, Vol. II, Somatosensory System, ed. A. Iggo. Springer-Verlag, Berlin, 1973. Figure 10; page 16.

Spinal Cord Lesion   Figure 15

A wound (or lesion) that severs one side of the spinal cord will interrupt both ascending and descending tracts. Degeneration of ascending tracts will occur above the lesion and degeneration of descending tracts will occur below the lesion (figure B). This is due to the cell body's functioning as the trophic* center of the neuron and all its processes, both axon and dendrites.

If an axon is cut or severed by some kind of lesion, the distal segment, that is, the part of the axon separated from the soma by the cut, degenerates. The severed distal segment can no longer maintain its structural integrity. The isolated axon breaks up into a series of membrane-bound spheres beginning near the point of damage and proceeding distally until the whole distal segment including the axonal endings has degenerated. The myelin likewise breaks down and is removed by scavenger cells. This degeneration from the point of damage distally is antegrade or Wallerian degeneration and invariably occurs to axons separated from their cell bodies both in the central nervous system as well as in peripheral nerves.

Locate neuron "1" in the lower left and follow its central process as it ascends to the level of lesion. The portion of fiber "1" above the lesion degenerates and this modality of sensation, discriminative touch (or active touch), which is carried up the ipsilateral dorsal columns, will be lost on the same side of the body below the level of the lesion (figure C).

Locate pain fiber "2" and trace it into the spinal cord where it synapses upon neuron "3" which crosses to the opposite side and ascends on the contralateral side. Above the level of the lesion the distal segment of neuron "3", separated from its trophic cell body, degenerates. Pain will be lost on the opposite side of the body below the level of the cut. Locate descending motor fiber "4" in figure B. Its cell body is at a higher level not shown in the drawing. Follow fiber "4" caudally to the lesion below which point its fiber degenerates.

Because of this, alpha motor neurons below the lesion are unable to receive motor commands from the brain and this part of the body is paralyzed. This type of paralysis affects the upper motor neuron but leaves the alpha motor neuron intact. Therefore, this type of paralysis is not characterized by an immediate flaccidity in voluntary muscles which follows damage to the alpha motor neuron or to its axon.

---

*Trophein (Greek) means "to nourish" or "sustain" and refers in this case to the apparent sustaining action or influence that the soma exerts upon its processes. Without this (chemical?) influence, the axons and dendrites degenerate. In addition to maintaining their component processes, neurons exert a trophic action upon nonnervous structures. Neural contiguity is essential for the structural integrity of sense organs such as taste buds as well as that of striated muscle. A neural trophic action has been strongly implicated in amphibian limb regeneration; a certain amount of nerve appears to be essential for an amputated newt limb to regenerate. If the nerves are removed from an adult newt limb, it will not regenerate following amputation.

The Cranial Nerves   Figure 16

Ventral aspect of brain stem showing exit of cranial nerves. There are twelve pairs of cranial nerves.

Cranial nerve I is the olfactory nerve (Latin, olfacia, "to smell"). It is made up of the central processes or axons of bipolar olfactory receptors in the top of the nose. These axons group into about twenty bundles which penetrate the cribriform plate of the ethmoid bone and end in the olfactory bulb. The axons of the olfactory nerve are thin and unmyelinated and are easily broken when the olfactory bulb is raised.

Cranial nerve II is the optic nerve. It consists of about one million fibers which arise from ganglion cells in the retina. The optic nerve is the only nerve that is enclosed by the three meninges, the dura mater, arachnoid mater, and pia mater. By convention the optic nerve is "nerve" from eyeball to chiasm. From chiasm to its end in the lateral geniculate body it is called "optic tract".

Cranial nerve III is the oculomotor nerve. It supplies motor fibers and movement to four extrabulbar eye muscles and the raiser of the eyelid (levator palpebrae).

Cranial nerve IV is the trochlear nerve (Greek, trochlea, "pulley"). It supplies only one extrabulbar eye muscle, the superior oblique. It is the only cranial nerve that be comes completely crossed (it supplies the opposite superior oblique). It is also the only cranial nerve that exits from the dorsal aspect of the brain stem.

Cranial nerve V is the trigeminal nerve (Latin, trigeminus, "three fold"). It is mainly sensory and divides into three large branches, the ophthalmic, maxillary, and mandibular, that supply the face, eye, nose, mouth, teeth, and anterior two-thirds of the tongue with pain, touch, and temperature sensibilities. It has a small motor component that supplies the muscles of the jaw (and a few others).

Cranial nerve VI is the abducent nerve (Latin, abducere, "to lead away"). It supplies only one extrabulbar eye muscle, the lateral rectus.

Cranial nerve VII is the facial nerve. It has two roots. Its larger motor root supplies the muscles of facial expression. Immediately lateral to the motor root is the smaller root of the facial nerve, the nervus intermedius. The nervus intermedius contains taste fibers from the anterior two-thirds of the tongue and parasympathetic fibers that supply the lacrimal gland, nasal gland, submandibular gland, and sublingual gland.

Cranial nerve VIII is the vestibulocochlear nerve. It carries two distinct sensory modalities from the inner ear. The older vestibular portion carries information about the position and movement of the head. The vestibular part of the inner ear responds to the head being turned, tilted, accelerated in a straight line. The newer cochlear portion of the vestibulocochlear nerve carries hearing from the cochlea.

Cranial nerve IX is the glossopharyngeal nerve (Greek, literally "tongue-throat"). It is mainly sensory to the upper pharynx and soft palate. It supplies pain, touch, temperature, and taste to the posterior one-third of tongue. It supplies motor fibers which join with those of the vagus nerve to form the pharyngeal plexus which innervates the soft palate and upper pharynx. Parasympathetic fibers in the glossopharyngeal nerve supply the parotid salivary gland.

Cranial nerve X is the vagus nerve. It contains five different kinds of fibers. These include preganglionic parasympathetic fibers that innervate the esophagus, heart, lungs, stomach, and intestines. There are numerous visceral afferent fibers that are concerned with cardiac and pulmonary reflexes. The vagus supplies motor control of the pharynx and larynx. Taste fibers from taste buds on the epiglottis travel within the vargus as do a few afferent fibers from the skin of the ear.

Cranial nerve XI is the accessory nerve (old name, spinal accessory nerve). It arises by two roots, a cranial one in the medulla and a spinal one from the cervical spinal cord. The spinal root ascends through the foramen magnum and joins the cranial root as the latter emerges from the medulla. It supplies only two muscles, the trapezius and sternocleidomastoid.

Cranial nerve XII is the hypoglossal nerve. It supplies all the tongue muscles on the same side.

Directions: Color each of the cranial nerves.

Cranial Nerve Components Figure 17

There are seven categories or components of fibers within the twelve cranial nerves. Some cranial nerves such as the vagus may contain five different components. Others such as the hypoglossal contain only a single component.

Directions: Use a different color for each of the components and related nuclei.

Figure A is a transverse section through the lower medulla and shows the relative position of the cranial nerve nuclei in terms of their components.

The three more medial motor nuclei contain the cells of origin of the three motor components (neurons "1", "2", "3"). The three more lateral sensory nuclei serve as terminal, integrative, and relay stations for the four sensory components. One nucleus, the nucleus of the solitary tract, receives the fibers of two components, thus there are seven components and only six nuclei at this level. Note that the cell bodies of the four sensory components lie outside the brain (in ganglia). These are afferent neurons "4", "5", "6", "7".

In figure A, locate neuron "1" in the GSE (general somatic efferent) nucleus. Four cranial nerves, III, IV, VI, and XII, carry GSE fibers and supply head muscles that are derived from the myomers of somites in the embryo. The only somatic (or myomeric) muscles in the face region are those of the tongue (nerve XII) and the extrinsic eye muscles (nerves III, IV, VI). Color neuron "1" and the GSE nucleus in figure A and, in figure B, the nuclei of nerves III, IV, VI, and XII.

Return to figure A and locate neuron "2" in SVE (special visceral efferent) nucleus. These motor neurons supply muscles that are derived from the branchial or gill arches and are called branchiomeric muscles (see figure 25). SVE fibers leave the brain stem in the following nerves: V, VII, IX, X, XI. Note the position of the SVE nucleus in relation to the GSE nucleus. Note also how the fiber of "2" first curves medially and dorsally before exiting as do the fibers of VII, IX, X before they exit. The SVE fibers all tend to exit more laterally on the brain stem than do the SVE fibers which exit ventrally (with the exception of nerve IV which exits dorsally). The perikarya of SVE motor neurons lie in the following SVE nuclei: motor nucleus of V (Mot Nuc V), facial motor nucleus (Nuc VII), the nucleus ambiguus (Nuc Ambig), and the spinal nucleus of XI (Sp Nuc XI).

Color neuron "2" in figure A and the above-mentioned SVE nuclei in figure B.

Return to figure A and locate neuron "3" in the GVE (general visceral efferent) nucleus which contains preganglionic parasympathetic cell bodies and represents the following GVE nuclei: Edinger-Westphal (EW) nucleus (new name, accessory oculomotor nucleus), superior salivatory nucleus (ss), inferior salivatory nucleus (is), and the dorsal nucleus of X (Dor Nuc X). The nerves containing preganglionic parasympathetic fibers are III, VII, IX, and X. Color GVE neuron "3" and each of the GVE nuclei.

GVA (general visceral afferent) includes conscious sensations from the pharynx and larynx which end in the caudal part of the solitary nucleus as well as subconscious impulses involved in a wide range of pulmonary, cardiovascular, and digestive reflexes that arise in the thorax and abdomen, and terminate in both the dorsal nucleus of X and in the nucleus of the solitary tract. GVA fibers travel within nerves IX and X and their somata lie in the inferior ganglia of both nerves.

Color GVA neuron "4" and GVA nucleus (nucleus of solitary tract) in figure A and the caudal part of the nucleus of the solitary tract (Nuc Solit Tr) in figure B.

Taste is SVA (special visceral afferent) and travels in three nerves, VII, IX, X. All taste fibers end centrally in the rostral part of the nucleus of the solitary tract. Color SVA neuron "5" and the SVA nucleus in figure A and the rostral part of the nucleus of the solitary tract in figure B.

Cutaneous face sensation and proprioception from the jaw muscles and the lips, cheek, and tongue musculature are designated GSA (general somatic afferent) and travel centrally within nerve V. GSA impulses end in the three sensory nuclei of nerve V, the mesencephalic nucleus of V (Mesen Nuc V), the principal (or main) sensory nucleus of V (Pr Sen Nuc V), and the spinal nucleus of V (Sp Nuc V). Color neuron "6", GSA nucleus, and the three sensory nuclei of V.

Fibers from the inner ear are SSA (special somatic afferent). These travel centrally within the VIII nerve. They end centrally in the four vestibular nuclei, the superior vestibular nucleus (sv), the medial vestibular nucleus (mv), the inferior vestibular nucleus (iv), and the lateral vestibular nucleus (lv); and in the two cochlear nuclei, the dorsal cochlear nucleus (d) and the ventral cochlear nucleus (v). Color the SSA neuron and the vestibular nuclei and cochlear nuclei.

Figure A   Based upon and modified from A. Brodal. Neurological Anatomy in Relation to Clinical Medicine, 3rd. ed. Oxford University Press, New York. 1981. Figure 7-1.

Figure B   Based upon and modified from Nieuwenhuys, Voogd, van Huijzen. The Human Central Nervous System. A Synopsis and Atlas. Springer Verlag, Berlin. 1978. Figure 93.

The Hypoglossal Nerve (XII), Accessory Nerve (XI), and Vagus Nerve (X)   Figure 18

On the right side, the drawing shows one neuron for each of the functional components carried in these three cranial nerves. On the left side, the nuclei associated with these nerves are depicted as longitudinal columns in the medulla. The drawing of the brain stem on the upper right indicates the level of the medulla through which the section passes.

The Hypoglossal Nerve (XII) supplies motor fibers to all the tongue muscles, both intrin-sic and extrinsic. Each nerve supplies the homolateral tongue musculature; thus, damage to this nerve will cause the tongue to deviate to the paralyzed side because of the lack of protrusor action of the paralyzed genioglossus muscle which will hold its side back. The motor neurons whose axons comprise nerve XII have their cell bodies in a longitudinal column called the hypoglossal nucleus. This nucleus lies in the lower medulla in a medial and dorsal position. The axons leave the nucleus, run ventrally to exit between the pyramids and inferior olive and group together to form the hypoglossal nerve. Each hypoglossal nucleus receives corticobulbar fibers (that is, fibers that run from the cerebral cortex to a nucleus either in medulla, pons, or midbrain) mainly from the opposite cerebral hemisphere, but the more medial parts of the nucleus receive fibers from both hemispheres (not shown in the drawing). The tongue musculature, along with the extraocular (voluntary) eye muscles, is derived from somites; thus the motor neurons and their axons that comprise the hypoglossal nerve are classified as general somatic efferent (GSE). The motor neurons of the hypoglossal nucleus are, as are all motor neurons of cranial nerves (excluding the parasympathetic neurons), "lower" or "peripheral motor neurons" since their axons terminate directly on striated muscle. Damage to their axons results in flaccid paralysis.

The Accessory Nerve (XI) has two origins, a cranial one in the nucleus ambiguus, and a spinal one in the accessory nucleus at levels C_2-C_5 in the spinal cord. The nucleus ambiguus lies lateral and somewhat ventral to the hypoglossal nucleus. However, it is not a compact cellular column as the drawing suggests, but a scattered collection of motor neurons, hence, its name the "ambiguous nucleus". The motor neurons that lie within the nucleus and their axons which exit with cranial nerves IX, X, and XI are classified as special visceral efferent (SVE), since they innervate the branchial muscles derived from branchial arches three, four, and six. Nerve XI axons arising from somata in the nucleus ambiguus exit from the medulla, join with those axons from the spinal component and form the accessory nerve. However, the fibers from the cranial component (nucleus ambiguus) join the vagus nerve, whereas those from the spinal component leave the skull, receive additional fibers from spinal nerves C_3 and C_4 and supply the sternocleidomastoid and trapezius muscles. Damage to the accessory nerve will result in an inability to fully elevate the arm since the trapezius muscle rotates the scapula when the arm is elevated beyond horizontal. Also there will be some difficulty in turning the head towards the unaffected side and in ventrally flexing the head from the supine position due to paralysis of the sternocleidomastoid muscle.

The Vagus Nerve (X) has five functional components, GSA, SVA, GVA, SVE, and GVE. It has two ganglia, a superior and an inferior, that contain the somata of its afferent fibers. Vagal GSA fibers carry cutaneous sensation from the skin of the eardrum, and end in the spinal nucleus of V. Vagal taste fibers (SVA) carry taste from taste buds on the epiglottis to the rostral part of the solitary nucleus. Some vagal GVA fibers carry sensation that reaches the conscious level. These include pain fibers from the mucosa of the larynx, lower pharynx, esophagus, stomach, and intestines. Other visceral sensations such as nausea and hunger also travel in the vagus. However, the great majority of visceral afferent fibers (GVA) within the vgus nerve operate at the subconscious level and play vital roles in cardiovascular, pulmonary, and digestive functions. These include fibers arising from chemoreceptors and pressor receptors in the great vessels that monitor the chemival composition and pressure of the blood, fibers from stretch receptors in the lung that respons to distention of the air passageways, and fibers from the walls of the digestive tract. Some vagal GVA fibers end in the solitary nucleus (shown in the drawing). Other vagal GVA fibers end in the dorsal nucleus of the vagus. The fiber labelled "col (GVA) IX X" represents a collateral from the carotid branch of nerve IX and carries impulses from either the carotid sinus, a pressor receptor, or the carotid body, a chemoreceptor.

Special visceral efferent fibers (SVE) arise from somata in the nucleus ambiguus. These motor neurons supply the intrinsic laryngeal musculature and most of the muscles of the pharynx. Note that some vagal SVE fibers are "transferees" from the accessory nerve. General visceral efferent fibers (GVE) are preganglionic parasympathetic fibers that have their secretomotor cells of origin in the dorsal nucleus of the vagus nerve (former name, dorsal motor nucleus of X). The fibers have a very widespread distribution throughout the thorax and abdomen where they supply the heart, lungs, stomach, and intestines. They synapse upon postganglionic parasympathetic neurons in ganglia near or within the target organ. Vagal parasympathetic fibers are responsible for the peristaltic wave in the esophageal smooth musculature, for gastric gland secretion and gastric motility, for relaxation of the pyloric sphincter and for intestinal motility.

Damage to the vagus nerve produces such symptoms as palpitations, increased pulse, constant vomiting, slowing of respiration, and a sense of suffocation. Because of vagal innervation of the eardrum, a plug of wax in the ear can produce an "ear cough", and syringing of the ear may produce vomiting and even cardiac inhibition. Unilateral paralysis of the soft palate will be noted when the patient says "ah". The normal side will be elevated, the paralyzed side will have a lower arch and the uvula will be pulled to the normal side. This is due to the unopposed action of the normal levator palatini. A permanently constricted pylorus may result from complete absence of vagal control.

Abbreviations: m lem, medial lemniscus; sol nuc, solitary nucleus; sp nuc V, spinal nucleus of V; sp tr V, spinal tract of V.

Directions: Color each of the functional components of nerves XII, XI, and X.

The Glossopharyngeal Nerve (IX) Figure 19

The <u>Glossopharyngeal Nerve</u> (IX), like the vagus nerve with which it is closely associated, has the same five functional components. It also has two ganglia, a superior and inferior, which contain the unipolar somata of its afferent fibers. It emerges from the medulla by several rootlets lateral to the olive and directly rostral and in line with the vagus and accessory nerves.

It contains the following functional components: A few special visceral efferent fibers (SVE) arising from the rostral part of the nucleus ambiguus and supplying the stylopharyngeus muscle; a few general visceral efferent (GVE) or, more specifically, preganglionic parasympathetic fibers originating from the inferior salivatory nucleus and synapsing in the otic ganglion where secretory impulses are relayed to the parotid gland; taste fibers or special visceral afferents (SVA) leading from taste buds on the posterior one-third of the tongue and the palatal arches and ending centrally in the rostral third of the solitary nucleus (these taste fibers have their somata in the inferior ganglion); a few general somatic afferent (GSA) fibers with somata in the superior ganglion conveying pain, touch, and temperature from the skin of the ear terminating in the spinal nucleus of V; and finally, general visceral afferent (GVA) fibers constitute the fifth group, some of which carry impulses that reach the level of consciousness and others remain at the subconscious level; the former group includes pain and temperature from the posterior one-third of the tongue and upper pharynx, and pain from the middle ear; the latter (subconscious) group includes afferents from the carotid sinus and carotid body.

All the visceral afferents have their somata in the inferior ganglion. There is evidence to indicate that some of the general visceral afferents in the glossopharyngeal nerve enter the solitary tract and end in the solitary nucleus. However, general visceral afferents from the posterior one-third of the tongue and the upper pharynx carrying pain and probably touch and temperature as well, or possible collaterals of these fibers, end in the spinal nucleus of V (not shown in drawing).

Afferents from the carotid body and the carotid sinus, or collaterals of fibers that end in the solitary nucleus (indicated by "col" in the drawing) end in the dorsal nucleus of X. Thus an increase in arterial pressure will be registered by the carotid sinus whose afferent fibers will cause the preganglionic parasympathetic neurons in the dorsal vagal nucleus to send inhibitory impulses via the vagus nerve to the heart (acetylcholine depresses the heartbeat) and the arterial pressure will be correspondingly lowered. The carotid body responds to hypoxia and brings about an increase in blood pressure, heart rate, and respiratory rate.

<u>Damage</u> to the glossopharyngeal nerve will result in a loss of all sensation from the posterior one-third of the tongue, the palatal arches, the palatal tonsil, and the upper pharynx. The "gag" reflex, which is an urge to retch or heave when the mucosa of the pharynx is touched, is lost on the afflicted side following damage to the glossopharyngeal nerve.

<u>Directions</u>: Using a different color for each functional component, color the soma and fiber of each of the five neurons that represent the five components of the glossopharyngeal nerve. The figure of the brain stem at the right and the transecting plane indicate the level of the section.

<u>Abbreviations</u>: inf vest nuc, inferior vestibular nucleus; inf olive, inferior

olivary nucleus; med vest nuc, medial vestibular nucleus; mlf, medial longitudinal fascicuus; sol tr, solitary tract; taste 2nd, second order taste relay neurons in the rostral solitary nucleus; col, collateral of a GVA fiber from the solitary nucleus to the dorsal vagal nucleus. Also see figure 18.

Central Vestibular Connections (N.VIII)   Figure 20

The vestibular portion of the vestibulocochlear nerve (N. VIII) consists of the central processes of bipolar vestibular ganglion neurons (1). The peripheral processes of the bipolar cells begin as synapses at five different sites: the hair cells in the three ampullary crests (2, crest of superior semicircular duct; 3, crest of lateral duct; 4, crest of posterior duct), the hair cells of the macula of the utricle (5), and the hair cells of the macula of the saccule (6). Fibers (7) that arise from the two maculae divide upon entering the brain stem into an ascending and descending branch with the ascending branch ending in the lateral vestibular nucleus (8) and the descending branch ending in the medial vestibular nucleus (9). Fibers that arise from the ampullary crests (10) also divide with one branch ending in the superior vestibular nucleus (11) and another in the rostral portion of the medial vestibular nucleus.

The ascending portion of the medial longitudinal fasciculus (MLF) arises from both the superior and medial vestibular nuclei, but not from the other two nuclei. The fibers that arise from the superior vestibular nucleus (12) remain essentially homolateral and supply the motor nuclei of nerves III and IV. However, they supply both the homolateral and contralateral nuclei III and IV by means of collaterals that cross the midline. The enlarged detail figure A (right) illustrates how these fibers end on both the homolateral as well as the contralateral nuclei III and IV. The superior vestibular nucleus may also send a minor contribution to the homolateral nucleus VI. Some fibers ascend to end in the homolateral interstitial nucleus of Cajal (13).

The rostral portion of the medial vestibular nucleus gives rise to MLF fibers that cross the midline and ascend on the opposite side. One set of fibers (14) crosses the midline, ascends on the opposite side where it bypasses nuclei VI and IV and supplies only nucleus III, where it supplies both sides. It then continues to the contralateral interstitial nucleus of Cajal. Another set of fibers (15) also crosses the midline and supplies only the contralateral (to the fibers' origin) nuclei VI and IV. Both sets of contralateral MLF fibers that are derived from the medial vestibular nucleus are shown more fully in figure B (lower right).

The MLF also descends into the spinal cord as far as the upper thoracic segments where it is also called the medial vestibulospinal tract. These descending MLF fibers arise mainly from the medial vestibular nucleus and descend on both sides of the spinal cord, contralateral (16) as well as homolateral (17). The lateral vestibulospinal tract (17) which is much larger and much longer, extending the full length of the cord, originates from neurons in the lateral vestibular nucleus. Unlike the medial vestibulospinal tract, the lateral vestibulospinal tract remains strictly homolateral. It projects to all levels of the spinal cord and ends mainly on interneurons and has a facilitatory action on extensor motor neurons, both alpha and gamma.

Abbreviations: III, oculomotor nucleus; IV, trochlear nucleus; VI, abducent nucleus; ml, midline.

Directions: Starting with fiber "7", trace the course of the vestibular fibers (7 and 10) centrally to their endings in the lateral, medial, and superior vestibular nuclei. Then trace the ascending MLF fibers, using one color for homolateral fibers (12) and another for the contralateral fibers (14 and 15). Do the same for the vestibulospinal tracts (16, 17, 18).

---
*Based mainly on Brodal (1981) and Tarlov (1970).

Brodal, A. Neurological Anatomy. Oxford University Press, New York. 1981.

Tarlov, E. 1970. Organization of Vestibulo-oculomotor Projections in the Cat. Brain Research 20, 159-179.

Central Auditory Pathways (N.VIII) Figure 21

The auditory (or cochlear nerve) portion of the vestibulocochlear nerve (N. VIII) consists of the central processes (1) of the bipolar neurons whose somata (25) comprise the spiral ganglia within the bony spiral lamina of the cochlea. (The cochlear nerve, in addition to afferent fibers, contains efferent fibers, the olivocochlear system, and sympathetic fibers.) The peripheral processes (6) of these neurons make synaptic contact with the inner (7) and outer hair cells (8) of the organ of Corti.

Sound vibrations in the perilymph cause the basilar membrane (9) to rise and fall which in turn causes a side-to-side (or radial) bending of the stereocilia (previously called "hairs") of the hair cells. The stereocilia (10) of the outer hair cells have their tips embedded in the overlying tectorial membrane (11). Recent evidence indicates that the stereocilia (12) of the inner hair cells are not attached to the tectorial membrane, as previously believed, but instead remain free and become deflected directly by movement in the surrounding endolymph.

Incoming impulses arising from the hair cells are brought to the pontomedullary junction, where each afferent fiber (1) divides into a descending dorsal branch and an ascending ventral branch which end in the dorsal (13) and ventral (14) cochlear nuclei respectively. The two cochlear nuclei project second-order fibers mainly to the region of the opposite superior olivary complex, which, in the cat, consists of lateral superior olive (15), a medial superior olive (16) and a more medial nucleus of the trapezoid body (17). The drawing of the central auditory pathways is based mainly upon the cat. In the human, the lateral superior olive and nucleus of the trapezoid body are poorly developed.

Not illustrated in the drawing are various connections between the superior olive complex and certain cranial nerve nuclei that subserve such reflexes as turning the head in re-sponse to sound (N. XI), pinnae orientation (pricking up of dog's ears) (N. VII), and turning the eyes (NN. VI, IV, III). Two small muscles within the middle ear, the stapedius (innervated by N. VII) and tensor tympani (innervated by N. V), contract in response to loud sounds and thus protectively dampen the vibration of the stapes and malleus, respectively. Recent work indicates that the stapedius routinely contracts during loud sound. The tensor tympani, on the other hand, only responds to extremely loud and painful sounds. These middle ear reflexes apparently protect the organ of Corti from excessive vibration and probably filter out noise generated in the head by chewing, talking, and singing.

The second-order fibers that arise from the two cochlear nuclei traverse the midline of the brain and converge upon the opposite superior olivary complex by three routes: a dorsal acoustic stria (18), an intermediate stria (not shown), and a much larger ventral acoustic stria or trapezoid body (19). Note that some second-order fibers (2d) end in the superior olivary complex whereas others turn rostrad (2a, 2b, 2c) and continue without interruption within a flattened band, the lateral lemniscus (20). Some neurons in the ventral cochlear nucleus project to the ipsilateral superior olive (2e), but not as many as project to the opposite superior olive (2d). Thus the central auditory pathways at the level of the lateral lemniscus and above carry impulses that are largely from the contralateral ear (about 60% contralateral, with 40% from the ipsilateral ear). Damage to one lateral lemniscus will result in a hearing impairment to both ears, but the opposite ear will be more severely affected.

Groups of somata along the course of the lateral lemniscus comprise the nucleus of the lateral lemniscus, most of which project up to the ipsilateral inferior colliculus (3e), but some project to the opposite inferior colliculus (3d). It was formerly believed that all ascending fibers ended in the nuclei of the inferior colliculus, however recent investigation in the chimpanzee has demonstrated second-order fibers extending from the cochlear nucleus to the contralateral medial geniculate body (2c). Up to the level of the inferior colliculi, the two auditory pathways freely communicate with each other by fibers that cross the midline. The commissure of the inferior colliculus is one such interconnection (3h).

Some neurons in the inferior colliculus are sensitive to the time interval that occurs when the sound strikes one ear before the other. The sound will reach the ear facing the sound source before it reaches the ear facing away from it. These neurons fire off in response to (for example) the right ear receiving the stimulus before the left. They are also able to detect the magnitude of this interval, thus allowing the animal to determine the direction of the sound. Other neurons in the inferior colliculus respond if the sound is louder at one ear than at the other. This, too, is a means of determining the source of the sound. The inferior colliculus appears to be a center where differences between the ears in respect to stimulus time and stimulus intensity are recorded; thus it plays an essential role in locating the origin of the sound in space.

Neurons in the inferior colliculus project upward (actually <u>upward and laterally</u>) to the medial geniculate body by a short tract, the brachium (Latin, arm) of the inferior colliculus (22). The fibers within the brachium of the inferior colliculus are mainly from the ipsilaterial inferior colliculus, but there is some projection from the contralateral inferior colliculus (3g). Neurons within the medial geniculate body (23), which is a nucleus of the thalamus, project by the auditory radiation (24) to the auditory cortex (25).

This area of the cortex lies on the superior surface of the superior temporal gyrus and is buried within the Sylvian fissure. One or two short transverse gyri, Herschl's gyri, mark the auditory cortical area which corresponds to area 41 of Brodmann. Usually the left side contains one longer Herschl gyrus and the right, two shorter ones. According to Brodmann's scheme, area 41 (Herschl's gyri) is the primary auditory cortex; surrounding this are areas 42 and 22, the auditory association cortex.

From animal studies, it is evident that the auditory cortex is essential for distinguishing different temporal patterns, such as one long tone followed by three short ones from three short tones followed by one long one. Bilateral ablation of the auditory cortex completely abolishes this ability in animals. Loudness appears to depend upon how many hair cells and neurons are stimulated and the rates at which they fire. The ability to determine different levels of loudness, or intensity discrimination, in animals is believed to reside in structures below the inferior colliculus.

In addition to the ascending auditory pathways just described, there is a descending auditory system that parallels the ascending one (not illustrated in drawing). It begins in the auditory cortex and ends in the opposite cochlea. Another feature of the central auditory pathways is that there is an orderly projection of fibers in respect to frequency from the cochlea to the auditory cortex with fibers that respond best to high frequencies at one position and those to low frequency at another position. This arrangement is referred to as a <u>tonotopic</u> (or cochleotopic) projection.

<u>Directions:</u> Using a different color for each order, color the central auditory pathways. 1 means first order; 2a-2e, second order; 3a-3i, third order, 4a-4b, fourth order; 5, fifth order.

Facial Nerve (VII) Figure 22

The facial nerve (VII) actually consists of two separate nerves: the larger facial nerve proper which is motor to the muscles of facial expression (mimetic muscles), and the smaller nervus intermedius which contains three types of fibers: taste, parasympathetic, and cutaneous sensation.

In terms of cranial nerve components, the motor fibers within the facial nerve proper that innervate the muscles of facial expression are SVE (the muscles of facial expression are branchiogenic, being derived from the second or hyoid arch); taste (from the anterior two-thirds of the tongue) is SVA; parasympathetic (to the submandibular, sublingual, palatal, nasal, and lacrimal glands) is GVE; and cutaneous sense (from the ear) is GSA.

The motor fibers to the muscles of facial expression (e.g. orbicularis oris, buccinator, platysma) arise from the facial motor nucleus which lies in the lower pons. In humans this is the largest motor nucleus of any cranial nerve and consists of 7,000-10,500 neurons; the motor (SVE) fibers in the facial nerve which arise from this nucleus are estimated to be about 7,000. Note that the motor fibers upon leaving the facial motor nucleus proceed medially and dorsally towards the fourth ventricle where they make a bend or <u>genu</u> around the abducent (VI) nucleus; they then turn ventrally and laterally to exit at the lower border of the pons.

The GVE fibers that travel in the nervus intermedius of VII are preganglionic parasympathetic fibers and arise from a small cluster of neurons, the superior salivatory nucleus. The preganglionic parasympathetic fibers synapse upon postganglionic parasympathetic neurons in the submandibular and pterygopalatine ganglia which relay secretory impulses to the glands mentioned above.

The taste fibers have their unipolar somata in the geniculate ganglion. Their peripheral processes end upon taste buds on the anterior two-thirds of the tongue. They travel centrally first within the lingual nerve. Usually they leave the lingual nerve by way of the chorda tympani nerve which connects the lingual nerve (a branch of V_3) with the facial. The central processes of the taste fibers enter the brain stem as part of the nervus intermedius and then end within the rostral portion of the solitary nucleus. The drawing shows the SVA (taste) fiber turning caudally where it will end in the solitary nucleus (not shown).

The few cutaneous fibers (GSA) also have unipolar somata in the geniculate ganglion. They convey general cutaneous sensation from the skin of the ear to the spinal tract of V and end in the spinal nucleus of V.

Damage to the motor fibers (SVE) of the facial nerve results in the peripheral facial nerve palsy (or Bell's palsy) in which the <u>ipsilateral</u> muscles of facial expression are affected to various degrees. If the paralysis is complete, all movement of the facial muscles is lost, flaccid paralysis is evident, the face shows a marked asymmetry, and unless the condition reverses itself, muscular atrophy ensues (see figure 24). Facial paralysis can also occur if there is damage to neural structures above the facial nucleus, such as a lesion to the cerebral cortex or the pathways that arise from the cortex and end in the facial nucleus (corticonuclear or corticobulbar pathways). Paralysis such as this is called supranuclear paralysis of the facial nerve (see figure 23).

<u>Directions</u>: Using a different color for each component, color and trace the course of the four components of the facial nerve.

Facial Nerve and Higher Motor Control Figure 23

Facial motor fibers arise from cell bodies in the facial motor nucleus (1). Notice that fibers (2) that innervate the upper facial muscles such as the frontalis muscle (3) arise from somata (4) that receive ipsilateral corticonuclear (also called corticobulbar) fibers (5) as well as contralateral corticonuclear fibers (6). Facial motor fibers (7) that innervate the lower facial muscles, however, such as the buccinator muscle (8), receive only contralateral corticonuclear fibers (9).

Therefore a lesion such as a cerebral hemorrhage in the internal capsule (10) could possibly destroy all corticospinal and corticonuclear fibers (the great majority of which end in the opposite side of the brain stem and spinal cord), thus paralyzing the opposite side of the body. Following such an accident, fibers innervating the upper facial muscles (2) would still function since they continue to receive some motor commands from the un-damaged ipsilateral corticonuclear fibers. Thus the eye could still be closed. However, motor neurons that give rise to fibers (7) innervating the lower facial muscles would receive no conscious motor control, since their corticonuclear fibers were destroyed in the capsular hemorrhage.

There is an additional "emotional pathway" from undetermined centers in the brain that supply the facial motor nucleus with "emotional motor commands" (as opposed to the voluntary motor commands from the cerebral cortex). These "emotional pathways" are believed to arise either from the globus pallidus (11) or from the hypothalamus (12).

Damage to the facial motor nucleus or to the facial nerve itself (13), such as occurs in Bell's palsy, would result in paralysis of the ipsilateral facial muscles. If the nerve were severed or destroyed at some point in its peripheral course, all the muscles of facial expression would be paralyzed, both upper as well as lower facial muscles. The facial motor neurons and their axons are lower motor neurons since their axons directly innervate voluntary muscle and their destruction results in instant flaccid paralysis and, if regeneration does not occur, eventual atrophy of the muscles of facial expression will happen.

In the case of destruction of corticonuclear fibers and the resultant paralysis of the contralateral lower facial muscles, genuine emotions such as laughter and sorrow are surprisingly expressed in the afflicted muscles, even though these same muscles remain unresponsive to conscious controls. In fact, the extent of the smile is greater, occurs earlier, and lasts longer on the affected side.

This observation, plus the loss of facial expression in diseases such as Parkinson's, suggests a dual control of the facial motor nucleus by higher motor centers, a voluntary one from the cerebral cortex and an "emotional" one mentioned above.

Directions: Using a different color for each of the following, color and trace the course of the facial nerve motor fibers (2 and 7), the homolateral corticonuclear fibers (5), the contralateral corticonuclear fibers (6 and 9), and the hypothetical "emotional" pathway originating from either "11" or "12".

Bell's Palsy   Figure 24

The most common disorder of the facial nerve is peripheral facial nerve paralysis or Bell's palsy. This is presumably due to some kind of inflammation or compression of the facial nerve in its course through the facial canal before its exit from the stylomastoid foramen. The causative agent has been attributed to viruses, vascular impingement against the nerve, and an abnormal bone growth compressing the nerve within the facial canal.

The affected muscles will be all the muscles of the facial expression (or mimetic muscles) plus the posterior belly of the digastric and the stylohyoid. The stapedius muscle of the middle ear receives a small branch from the facial nerve in the upper part of the facial canal and will be spared if the nerve blockage is below the origin of its nerve.

If the paralysis is complete all movement of the muscles of the facial expression will be lost on the same side. The face will be asymmetrical due to the muscles on the normal side pulling on the affected side. The affected side will be smooth and common facial features such as the nasolabial fold will be lost. The patient will be unable to frown or to close the eye due to paralysis of the orbicularis oculi. In fact, the palpebral fis-sure (area of the open eye) will be even greater than normal. The cornea may be in danger of drying and ulceration.

The mouth cannot be closed properly. The affected corner of the mouth (oral angle) is lower than normal with saliva sometimes drooling from it.

Since this is a type of peripheral or lower motor neuron paralysis, the affected facial muscles will be flabby and, if regeneration of the facial nerve does not occur, the paralyzed muscles will atrophy (waste away).

The affected muscles will not partake in any of the normal facial movements such as speaking, weeping, or eating, and food will tend to accumulate between the teeth and the paralyzed buccinator muscle.

The onset of Bell's palsy is usually sudden. The paralysis may only last for a few hours and then disappear. Eighty percent of the patients tend to recover within a few weeks.

Unlike supranuclear facial palsy in which the upper facial muscles such as the orbicularis oculi and frontalis retain their function, peripheral facial nerve paralysis involves all the ipsilateral mimetic muscles.

Figure 24 drawn by Joseph Kanasz.

The Trigeminal Nerve (V) and Central Connections   Figure 25

The trigeminal nerve (V) is a mixed nerve with a large sensory component (GSA) and a small motor component (SVE). It is estimated that the V nerve contains about 140,000 sensory fibers and 8,100 motor fibers.

The trigeminal nerve has three large branches ("trigeminus" in Latin means "three-fold" or "triplet"), the ophthalmic (V_1), the maxillary (V_2), and the mandibular (V_3). The ophthalmic supplies sensation to the upper face including the forehead and eye. The maxillary supplies sensation to the middle face including the upper lips, most of the nose, the nasal mucosa, the hard palate, and the upper teeth. The third division, the mandibular, supplies sensation to the lower face including the lower lips, the lower teeth, the mucosa of the cheeks and the anterior two-thirds of the tongue (pain, touch, and temperature only, taste is conveyed by the VII nerve).

The motor fibers of the V nerve travel with mandibular nerve and supply all the muscles derived from the first branchial arch (see figure 26). These are the four muscles of mastication (the temporalis, masseter, lateral pterygoid, medial pterygoid), the mylohyoid, the anterior belly of the digastric, the tensor tympani, and the tensor veli palatini.

Most of the sensory fibers have their perikarya in the semilunar ganglion (semil gang). These are typical unipolar neurons similar to those in dorsal root spinal ganglia. Their incoming central processes end in one of three sensory nuclei that receive trigeminal sensory fibers, the trigeminal mesencephalic nucleus (or mesencephalic nucleus of V; mes nuc V in drawing), the trigeminal main sensory nucleus (or main sensory nucleus of V; m sen nuc V), the trigeminal spinal nucleus (or spinal nucleus of V; sp nuc V) which in turn is divided into three parts, the upper nucleus oralis (oralis), below that, the nucleus interpolaris (inter), and below that, the lowest of the three, the nucleus caudalis (caud).

Locate fiber number "1" which is a large-diameter afferent fiber carrying discriminative touch or proprioceptive information from the mouth, tongue, or lips. Its unipolar perikaryon lies in the semilunar ganglion and its central process terminates upon neuron "2" in the main sensory nucleus which gives rise to an ascending fiber that crosses the midline and joins the ventral trigeminal (or trigeminothalamic) tract (V trig tr) which ascends to the ventralis posteromedialis (VPM) nucleus of the thalamus where neuron "3" relays the impulse to the "face" region of the somatosensory cortex.

The main sensory nucleus also gives rise to another ascending bundle of fibers, the dorsal trigeminal (or trigeminothalamic) tract (d trig tr in drawing). This tract arises from the dorsomedial portion of the main sensory nucleus (4) and, unlike the ventral trigeminal tract, the dorsal trigeminal tract remains entirely ipsilateral. Its function is not understood. It arises from a part of the main sensory nucleus that is concerned with the mouth and terminates in its own region of nucleus VPM (5). It may possibly be involved in taste.

Locate fiber "6" which is a pain fiber traveling with the mandibular nerve and carrying pain from one of the teeth in the lower jaw. Note that its unipolar perikaryon is in the semilunar ganglion and that its central process turns caudally and descends within a bundle of fibers, the spinal trigeminal tract (or spinal tract of V; sp tr V in drawing). It descends to the lowest of the three nuclei that comprise the spinal trigeminal nucleus, the nucleus caudalis (caud) where it ends by synapsing upon second order neuron "7" that projects its fiber across the midline where it joins the ventral trigeminal tract in its course to the opposite VPM nucleus. Third order neuron "8" in the VPM then projects this information to the somatosensory "face" cortex.

The nucleus caudalis appears to process pain and temperature data from the face, mouth, and nose. It resembles the dorsal horn of the spinal cord with which it blends at about the third cervical level. The spinal tract of V likewise blends with the dorsolateral fasciculus (tract of Lissauer) which carries pain and temperature fibers from the body.

There is another pathway through which pain from the face reaches the thalamus; this is a multineural network within the reticular formation (9). It has been proposed that the ventral trigeminal tract carries the sharp, well-defined pain similar to that carried in the lateral spinothalamic tract, while the reticular formation carries the dull, diffuse aching pain similar to that carried by the spinoreticular tract in the spinal cord.

Return to the right side and locate fiber "10" which carries touch. (This is probably the simple or crude touch mediated by bare nerve endings and falling within the "protopathic" designation of Head.) Note that it bifurcates upon entering the brain stem with one branch going to the main sensory nucleus and the other to the nucleus oralis. Second order neuron "11" relays this tactile information via the ventral trigeminal tract to the opposite VPM nucleus.

Nerves VII, IX, and X also contribute fibers to the spinal trigeminal tract. These three nerves carry general sensibilities from the external and middle (?) ear. Their central processes descend in the spinal trigeminal tract and end in the spinal nucleus (shown cut off on the left side).

Locate fiber "12" (on both sides). It is a first order sensory fiber, but is unlike any others because its cell body is not in the semilunar ganglion, but rather in the trigeminal mesencephalic nucleus (mes nuc V). This is the only case of a first order sensory (most likely proprioceptive) perikaryon lying within the CNS instead of within a ganglion. These fibers form a small but well-defined tract, the trigeminal mesencephalic tract (or mesencephalic root of V; mes tr V) which lies directly lateral to the mesencephalic nucleus of V. The mesencephalic nucleus of V is believed to process proprioceptive information that controls the force of the bite.

The fibers within the mesencephalic tract of V have been shown to arise from proprioceptors in the jaw muscles, the temporo mandibular joint, and the periodontal membrane of the teeth. Some collaterals from the afferent fibers in the mesencephalic tract of V project to the trigeminal motor nucleus (or motor nucleus of V; mot nuc V) where they synapse directly upon the motor neurons (13) within this nucleus.

Axons arising from these motor neurons form the motor component of the V nerve. They travel with the mandibular nerve and supply motor control to the muscles mentioned earlier. This two-neuron arc (12 and 13) between the sensory neurons of the mesencephalic nucleus of V and the motor neurons of the motor nucleus of V forms the basis for the monosynaptic jaw reflex in which a downward tap on the chin elicits a muscular contraction of the jaw muscles similar to the knee jerk.

Neuron "14" is a mesencephalic neuron whose collateral projects to the cerebellum (cut in drawing). (The collaterals of the neurons in the mesencephalic nucleus are in effect central processes since they carry the impulse to the target neurons.)

On the left, fiber "15" carries touch from the conjunctiva of the eye. Its central process enters the spinal tract of V and ends upon a second-order neuron in the nucleus oralis which relays this to motor neurons (16) in both facial motor nuclei (nuc VII) which in turn reflexly send motor commands, via nerve VII, to both orbicularis oculi muscles to blink. (The levator palpebrae, innervated by nerve III, would have to be simultaneously inhibited for the lid to close.) Fiber "17" on the left represents a pain fiber from the eye.

Directions: Color the listed nerve fibers using a different color for each functional group.

The Branchial Arches and The Branchiomeric Muscles Figure 26

The upper left figure A shows a human embryo at 32 days, at which time four branchial (Greek, branchia, gill) arches appear on the sides of the head and neck. The numbers 1, 2, 3, 4 indicate the four arches and the Roman numerals V, VII, IX, X are the four cranial nerves that supply motor innervation to the muscles of the four arches. Actually, there are six arches; but arch five rapidly disappears leaving no derivatives, and arch six does not form a definitive arch that is visible externally. However, arch six gives rise to most of the muscles of the larynx. Color nerves V, VII, IX, X in figure A.

Figure B shows a horizontal section through the branchial arches, each with its own cranial nerve. Color each of these cranial nerves (V, VII, IX, X).

Figure C shows a portion of the developing brain with the motor nuclei that contain the motor neuron cell bodies that supply the branchial muscles. Branchial arch one is the mandibular arch. It is supplied by the trigeminal nerve (V). The motor neuron perikarya lie in the motor nucleus of V and supply all the muscles that come from this arch; these are the muscles of mastication, which include the masseter, temporalis, medial pterygoid, and lateral pterygoid. Also derived from the first arch and supplied by the V nerve are the mylohyoid, anterior belly of the digastric, tensor veli palatini, and tensor tympani.

Figure D shows the temporalis and masseter muscles. Color nerve V starting with its perikaryon in motor nucleus of V (also called trigeminal motor nucleus).

The second arch is the hyoid arch and is supplied by the facial nerve (VII). Starting with the motor nucleus of VII (or the facial motor nucleus), color the course of a motor neuron of nerve VII. Note that it makes a loop before leaving the brain. The second branchial arch gives rise to all the muscles of facial expression. These are relatively small muscles that often insert into the skin. Some are essential such as the orbicularis oculi that protects the eye. The orbicularis oris and buccinator are used in normal chewing and sucking. In addition to supplying motor control to the muscles of facial expression, the facial nerve also supplies the posterior belly of the digastric, the stylohyoid, and the stapedius.

The third arch gives rise to the upper pharynx, and its nerve is the glossopharyngeal (IX). The IX nerve carries sensation from the posterior tongue and upper pharynx, and one would expect that the muscles of the upper pharynx, such as the superior constrictor and muscles of the soft palate, would also be supplied by IX. But the only muscle sup-plied by IX has been reported to be the stylopharyngeus (asterisk in figure F). Color and trace the course of nerve IX. Note that its soma lies in the rostral part of the nucleus ambiguus and also takes curved course in the brain stem.

The fourth arch develops into the muscles of the lower pharynx which are supplied by the vagus nerve (figure G). The muscles of the larynx actually develop from the sixth arch (not shown) and are supplied by the recurrent laryngeal branch of nerve X. Trace the course of nerve X and note that its cell body lies in the caudal portion of the nucleus ambiguus.

The muscles that are derived from the branchial arches are called branchiomeric muscles. These muscles are striated and voluntary and histologically identical to muscles that develop from somites (somatic or myomeric muscles). The only apparent difference is their embryonic origin. The nerves that supply these branchiomeric (branchiogenic would actually be a better name) are called special visceral efferent or simply SVE.

Each of these four cranial nerves has one or two ganglia which contain the sensory cell bodies of its sensory fibers which may greatly outnumber the motor fibers. These ganglia (figure C) are the semilunar ganglion (gg V) of nerve V, the geniculate ganglion (gg VII) for nerve VII, the superior and inferior ganglion (gg IX) for nerve IX, and the superior and inferior ganglion (gg X) for nerve X. The motor fibers in these four cranial nerves pass through these ganglia with no interruption. In fact, none of the ganglia contain any synapses, only unipolar perikarya of sensory fibers.

The Autonomic Nervous System   Figure 27

The autonomic nervous system oversees our fundamental visceral processes such as digestion, assimilation, the flow of blood, and the rate and amplitude of the heartbeat. It regulates the secretion of glands, the contraction of smooth muscle, the caliber of blood vessels, and cardiac output. Since it supplies secretomotor fibers to the viscera of the thorax and abdomen, it is also referred to as the visceral nervous system. It works as an involuntary system, reflex in nature, essentially at the unconscious level.

General plan. Figure A. The autonomic nervous system (ANS) consists of two antagonistic secretomotor systems, the parasympathetic and sympathetic. These have opposing actions on the organs innervated. Not all organs, however, receive this double innervation; for example, the sweat glands, arrector pili muscles, and blood vessels receive only sympathetic fibers. The autonomic outflow, in contrast to the somatic efferents, consist of a two-neuron chain. Thus, unlike axons of alpha and gamma motor neurons that extend from cell bodies in the anterior horn directly to striated muscle, the ANS efferents consist of two neurons in tandem. The first neuron's cell body is either in the brain stem or spinal cord. The second neuron's soma is in a ganglion somewhere outside the CNS. The first neuron is the preganglionic neuron, and the second, the postganglionic neuron. The axon of the first (preganglionic) neuron synapses upon the soma and dendrites of the second (postganglionic) inside the ganglion.

Parasympathetic preganglionic somata reside in four nuclei in the brain stem, the Edinger-Westphal nucleus (EW nucleus), the superior salivatory nucleus (sup sal nucleus), the inferior salivatory nucleus (inf sal nucleus), and the dorsal vagal nucleus (dorsal nucleus X). The preganglionic parasympathetic fibers arising from neurons in these ganglia leave the brain stem with cranial nerves III, VII, IX, and X, respectively. Color preganglionic parasympathetic neurons "1"-"5".

The postganglionic parasympathetic somata reside in four ganglia in the head and receive the terminal synapses of preganglionic fibers in the following manner: Preganglionic fibers in nerve III synapse upon postganglionic neurons in the ciliary ganglion; preganglionic fibers in nerve VII end upon postganglionic neurons in two ganglia, the pterygopalatine and submandibular; preganglionic fibers in nerve IX synapse upon postganglionic neurons in the otic ganglion, and preganglionic fibers in nerve X make synaptic contact with postganglionic neurons in numerous ganglia spread throughout the thorax and abdomen. Color postganglionic parasympathetic neurons "7"-"12".

Additional parasympathetic preganglionic fibers arise from somata in the sacral spinal cord and emerge with spinal nerves S_2-S_4. Since the parasympathetic system arises from four cranial nerves and three sacral nerves, it is also referred to as the craniosacral division. Color preganglionic parasympathetic neuron "6" in the lower spinal cord and postganglionic parasympathetic neuron "13" near the bladder.

The sympathetic preganglionic somata lie in the gray matter of the spinal cord at levels T_1-L_2 and emerge with the ventral roots of the corresponding spinal nerves; thus it is sometimes called the thoracolumbar division (or system). Postganglionic sympathetic somata are found in ganglia of the sympathetic chain (or paravertebral ganglia). Color preganglionic sypathetic neurons "14"-"17".

Additional postganglionic sympathetic somata reside in prevertebral ganglia, such as the celiac ganglia. As mentioned above, preganglionic sympathetic fibers leave the spinal cord in ventral roots and enter the spinal nerves. Once in the spinal nerves, they turn off in short white communicating rami (or branches) to enter the sympathetic trunk and ganglia. The preganglionic fibers may synapse upon postganglionic neurons in these ganglia; or they may ascend or descend in the trunk to end at a ganglia higher or lower; or they may continue into the thorax and abdomen without synapsing as splanchnic nerves. Color postganglionic sympathetic neurons "18"-"20".

Postganglionic fibers arising from somata in the ganglia of the sympathetic chain may either join spinal nerves via gray communicating rami or pass into the arms and legs with blood vessels.

Cholinergic versus adrenergic. Figure B. These terms refer to the mode of action of the two systems, specifically, the neurotransmitter released at the terminals of the postganglionic fibers. The postganglionic parasympathetics release acetylcholine (Ach); thus, it is a cholinergic system. The postganglionic sympathetics release norepinephrine (NE) (same as noradrenalin); thus, it is an adrenergic system. Exceptions to this are the postganglionic sympathetic fibers to sweat glands (sudomotor) which are cholinergic.

Parasympathetic (cholinergic) system is concerned with the vegetative aspects of day-to-day living. It promotes the digestion and absorption of food, gastric secretion, the motility of the intestinal tract (peristalsis), the relaxation of the pyloric and urinary sphincters, and slows the heart. Its action tends to be limited and discrete. This is due to each parasympathetic preganglionic neuron synapsing on relatively few postganglionic neurons (the reverse is true of the sympathetic system), and acetylcholine is rapidly broken down by the enzyme cholinesterase. The parasympathetic system promotes the acquisition and storing of free energy within the organism.

Sympathetic (adrenergic) system accelerates the heart (tachycardia) and raises the blood pressure by constricting blood vessels in the viscera and skin (which also limits heat loss from the skin). It dilates the bronchioles by relaxing the smooth muscles in their walls and inhibits peristalsis of the intestines. It causes the breakdown of glycogen into glucose and the liberation of free fatty acids, both of which supply more energy. The adrenergic action may be significantly augmented by the liberation of epinephrine from the adrenal medulla into the bloodstream. The cells of the adrenal medulla receive preganglionic sympathetic fibers in the splanchnic nerves. The complementary action of the adrenal medulla's hormones in the bloodstream, plus one preganglionic fiber exciting a large number of postganglionic fibers, plus norepinephrine's not being broken down as readily as acetylcholine, account for the adrenergic effect being massive and unitary in time of stress.

Some generalizations. There are no sympathetic cell bodies in the head;* sympathetic fibers in the head arise from cell bodies in the superior cervical ganglion. Parasympathetic fibers do not extend into the arms or legs, nor do they run along with or innervate blood vessels. All preganglionic neurons, parasympathetic and sympathetic, are cholinergic. Some postganglionic sympathetic neurons are cholinergic (sudomotor), the remaining postganglionic sympathetic fibers are adrenergic. The apparent contradictory action of noradrenalin (both excitatory and inhibitory) is believed to be due to there existing two types of receptors to the same neurotransmitter; one type of receptor will cause an excitatory action; the other, an inhibitory action.

Figure B shows some of the differences between the two divisions: Parasympathetic preganglionic fibers are longer and tend to synapse with fewer postganglionic neurons which are near or inside the target organ. Sympathetic preganglionic fibers are not as long and synapse upon many postganglionic neurons. All preganglionic neurons, parasympathetic and sympathetic, release acetylcholine (Ach); all postganglionic parasympathetic fibers release acetylcholine (Ach). Most postganglionic sympathetic neurons release norepinephrine (NE), but some release acetylcholine (Ach).

Directions: Color each of the four types of autonomic fibers.

---

*Some postganglionic sympathetic cell bodies are found in small groups along the internal carotid artery. These have apparently migrated cephalad from the superior cervical ganglion.

Basic Pathways of the Autonomic Nervous System Figure 28

The drawing shows a small portion of the thoracic spinal cord and related autonomic neurons. All preganglionic sympathetic perikarya are locatd in the intermediolateral column of the spinal cord gray matter. The intermediolateral column extends from the first thoracic level (T_1) to the second lumbar level (L_2) of the spinal cord. The axons of these preganglionic sympathetic neurons leave the spinal cord within the ventral roots of spinal nerves T_1 through L_2. Thus the sympathetic division leaves the CNS within the twelve thoracic nerves and upper two lumbar nerves. The parasympathetic division, on the other hand, leaves the CNS within four cranial nerves (see figure 29) and sacral nerves S_2, S_3, S_4.

Directions: Starting with neuron "1a", a preganglionic sympathetic neuron lying in the intermediolateral column of the spinal cord, color its cell body and axon as it passes out of the ventral root into the spinal nerve and then via a white communicating ramus into a ganglion of the sympathetic trunk where it ends by synapsing upon postganglionic sympathetic neuron "2c". Using another color for postganglionic sympathetic neurons, color neuron "2c" and its fiber as it runs via a gray communicating ramus back into the spinal nerve. Do the same for neuron "1b" and follow its axon which ascends in the sympathetic trunk to the superior cervical ganglion where it synapses upon postganglionic sympathetic neurons "2a" and "2b".

Color "2a" and "2b" and notice that the axons of "2a" form a plexus around the internal carotid artery. These postganglionic sympathetic axons ascend into the head as the internal carotid plexus, eventually reaching organs such as the eye and salivary glands.

Postganglionic sympathetic neurons "2b" have axons that innervate the heart via the cardiac plexus. Follow and color the course of preganglionic fiber "1c". Notice that "1c" passes through the sympathetic ganglion without synapsing and does not end until it reaches a ganglion such as the celiac ganglion. The thoracic splanchnic nerves are comprised of preganglionic axons such as "1c".

In the celiac ganglion, preganglionic fibers synapse upon postganglionic neurons "2h", "2i". Postganglionic neuron "2h" directs its axon to a blood vessel on the intestine. Postganglionic neuron "2i" sends its axon into the smooth muscle of the intestine. Follow the course of postganglionic sympathetic neurons "2d", "2e", "2f" whose cell bodies lie in another ganglion of the sympathetic trunk. Their axons either run into spinal nerves or run as a plexus around blood vessels and eventually end on such structures as sweat glands, arrector pili muscles, and blood vessels of the skin.

Locate preganglionic sympathetic neuron "1e" in the opposite intermediolateral column and trace its axon. Note that the axon of "1e" does not synapse in the sympathetic chain ganglia but continues into the abdomen as a splanchnic nerve. It synapses upon cells that comprise the adrenal medulla (2g) and secrete epinephrine (adrenalin), and lesser amounts of norepinephrine (noradrenalin) into blood vessels passing through the gland.

Neuron "3" represents a preganglionic parasympathetic neuronal cell body lying in the dorsal nucleus of the vagus nerve. Axons from somata such as "3" travel within the vagus nerve and supply a very wide range of viscera in both the thorax and abdomen.

The preganglionic parasympathetic fibers tend to be much longer than the postganglionic parasympathetic fibers.

Neurons labelled "4" are postganglionic parasympathetic neurons. Note that they lie entirely within the structure innervated. Postganglionic parasympathetic somata found in this position are therefore termed <u>intramural</u> (Latin, within the walls).

Neuron number "5" represents <u>visceral afferent neuron</u>. Its cell body lies in the dorsal root ganglion and has the typical unipolar shape of all the perikarya found in the dorsal root ganglion.

Parasympathetic Outflow in the Head   Figure 29

Preganglionic parasympathetic somata lie within four nuclei in the brain stem, the accessory oculomotor nucleus* (1), which is part of the oculomotor nucleus, the superior salivatory nucleus (2), the inferior salivatory nucleus (3), both in the lower pons, and the dorsal vagal nucleus** (4). The somata in these four nuclei give rise to preganglionic parasympathetic fibers that leave the brain stem with four cranial nerves (in respective order):   nerve III, nerve VII, nerve IX, and nerve X.

It should be pointed out that the preganglionic parasympathetic fibers are only one component of these four cranial nerves and comprise a minor portion in each of these nerves. The vagus nerve, for instance, contains five different kinds of fibers of which the parasympathetics are just one component.

These preganglionic parasympathetic fibers synapse upon postganglionic parasympathetic somata in four ganglia: Those of the oculomotor nerve end upon somata in the ciliary ganglia (5); those of the facial nerve upon somata in two ganglia, the pterygopalatine ganglion (6) and the submandibular ganglion (8); and those in the glossopharyngeal nerve upon somata in the otic ganglion (7).

Preganglionic fibers within the vagus travel considerable distances before synapsing upon postganglionic somata in organs such as the lungs, heart, stomach, and intestines (23).

Postganglionic fibers in the head arise from somata with the four above-mentioned ganglia and travel to the eye, lacrimal gland, nasal glands, palatine glands, submandibular gland, sublingual gland, and parotid gland.

Directions: Use one color for all of the preganglionic neurons (9, 10, 13, 14, 20, 22), and another for all the postganglionic neurons (11, 12, 15, 16, 17, 18, 19, 21, 23). After coloring all of the neurons, color the four parasympathetic nuclei (1, 2, 3, 4) another color, and the four parasympathetic ganglia (5, 6, 7, 8) still another color.

Start with neurons "9" and "10" in the accessory oculomotor nucleus. They both leave the mesencephalon as part of the oculomotor nerve (N III) which enters the orbit through the superior orbital fissure. They then leave the oculomotor nerve and enter the small ciliary ganglion where they end by synapsing upon postganglionic neurons "11" and "12". Neuron "11" innervates the constrictor (circular) muscle of the iris which reduces the size of the pupil diameter (miosis, Greek, "a lessening"), in this way it protects the eye from too much light. Postganglionic fiber "12" reaches the ciliary body where it causes the circular ciliary muscle to contract. This in turn allows the lens to assume a greater curvature and brings about focusing for near objects.

The two top figures show the lens shape for far vision (L_1) and for near vision (L_2). The lens is held in a relatively flat shape by the pull of the suspensory ligament (or zonules) exerted upon its periphery. When the circular muscle in the ciliary body contracts, this tension is reduced and the lens bulges into a somewhat more spherical shape which is its natural tendency. This tendency of the lens to assume a greater curvature is lost as the lens grows older and thicker (but not rounder) resulting in a form of far-sightedness due to old age, presbyopia (Greek, "old eyes").

Return to the brain stem and locate the superior salivatory nucleus (2). The fibers of the two preganglionic neurons "13" and "14" in the superior salivatory nucleus reach their respective ganglia by two quite different routes. Neuron "13" reaches its destination, the pterygopalatine ganglion, by way of the "greater petrosal route", and neuron "14", the submandibular ganglion, by way of the "chorda tympani route". Trace neuron "13" as it leaves the superior salivatory nucleus with the facial nerve. It passes through the geniculate ganglion (Gg VII) at which point it leaves the facial nerve and runs through a thin bony canal as the greater petrosal nerve. It emerges from the canal in the middle cranial fossa, then exits the skull through the foramen lacerum. It is then joined by the deep petrosal nerve (28) with which it traverses the pterygoid canal (28). The deep petrosal nerve consists of postganglionic sympathetic fibers derived from somata in the superior cervical ganglia and scattered somata along the internal carotid artery. Within the pterygoid canal the greater petrosal nerve and deep petrosal nerve unite to form a single nerve, the nerve of the pterygoid canal. Emerging from the canal, the greater petrosal nerve fibers end by synapsing upon postganglionic parasympathetic somata (15, 16, 17) within the pterygopalatine ganglion (6). The sympathetic fibers of the deep petrosal nerve merely pass through the ganglion without interruption.

Postganglionic neuron "15" supplies secretory fibers to the lacrimal gland. Note that fiber "15" enters the maxillary nerve (V_2), then continues to its zygomaticotemporal branch (29) which enters the orbit. It leaves the zygomaticotemporal branch by a thin communicating branch to the lacrimal nerve (26) which is a branch of the ophthalmic nerve (V_1). Fiber "15" continues within the lacrimal nerve until it reaches the lacrimal gland. The two postganglionic neurons "16" supply secretory fibers to nasal glands (30), and neuron "17" supplies secretory fibers to palatine glands (31).

Return to the superior salivatory nucleus and trace the course of neuron "14". Note that "14" also passes through the geniculate ganglion. But unlike the greater petrosal nerve (13) which leaves nerve VII at the geniculate ganglion, "14" curves downward with nerve VII in its course through the facial canal. Within the facial canal, "14" turns off from the facial nerve and becomes the chorda tympani nerve which travels up though its own canal, then into the middle ear cavity where it runs between the malleus and incus, and can sometimes be easily seen in a living subject with an otoscope. The chorda tympani (14) then leaves the skull through the petrotympanic fissure (a small crack in the mandibular fossa of the temporal bone), enters the back of the lingual nerve (33) which is a branch of the mandibular nerve (V_3), and travels with the lingual nerve into the tongue. Fiber "14" ends in the submandibular ganglion (8) whose postganglionic neurons (18, 19) innervate the sublingual (35) and submandibular (36) ganglia respectively.

Return to the brain stem and locate neuron "20" in the inferior salivatory nucleus which represents the parasympathetic component of the glossopharnygeal nerve (N IX) and is destined for the parotid gland. Follow the course of "20" and notice that it leaves the glossopharyngeal nerve (N IX) between its two ganglia which lie in the jugular foramen. This branch of nerve IX is the tympanic nerve and contains both sensory fibers and parasympathetic fibers. The tympanic nerve enters the middle ear cavity where the sensory fibers ramify to form the tympanic plexus (asterisk). The parasympathetic fibers, however, merely pass through the plexus, and form the lesser petrosal nerve that follows a bony tunnel near that of the greater petrosal nerve. The lesser petrosal nerve exits the skull through the foramen ovale and ends in the otic ganglion (7) upon postganglion neurons (21).

Postganglionic fiber "21" leaves the otic ganglion and "hitch-hikes" upon another branch of the mandibular nerve, the auriculotemporal (32), in which it rides until it reaches the parotid gland.

Return once again to the brain stem and locate the dorsal vagal nucleus (4) and its pre ganglionic parasympathetic neuron (22). The dorsal vagal nucleus is by far the largest of the four parasympathetic nuclei. The axons of its preganglionic neurons flow out with the vagus nerve (N X) and have a very extensive distribution to the lungs, heart, stomach, and intestine.

Postganglionic parasympathetic fibers tend to be much shorter than preganglionic parasympathetic fibers and are often located entirely (both soma and fiber) within the organ innervated. These intramural (Latin, "within the walls") ganglia are often seen in microscopic sections of the intestines.

---

*Old name: Edinger-Westphal nucleus.
**Old name: Dorsal motor nucleus of vagus nerve.

1. Accessory oculomotor nucleus
2. Superior salivatory nucleus
3. Inferior salivatory nucleus
4. Dorsal vagal nucleus
5. Ciliary ganglion
6. Pterygopalatine ganglion
7. Otic ganglion
8. Submandibular ganglion

9,10. Preganglionic parasympathetic neurons of N III
11,12. Postganglionic parasympathetic neurons of ciliary ganglion
13,14. Preganglionic parasympathetic neurons of superior salivatory nucleus
15,16,17. Postganglionic parasympathetic neurons of pterygopalatine ganglion
18,19. Postganglionic parasympathetic neurons of submandibular nerve
20. Preganglionic parasympathetic neuron of inferior salivatory nucleus
21. Postganglionic parasympathetic neuron of otic ganglion
22. Preganglionic parasympathetic neuron of dorsal vagal nucleus
23. Postganglionic parasympathetic neuron in heart, intestine, etc.

24. Iris
25. Ciliary body
26. Lacrimal nerve
27. Pterygoid canal
28. Deep petrosal nerve
29. Zygomaticotemporal nerve
30. Nasal glands
31. Palatine glands
32. Auriculotemporal nerve
33. Lingual nerve
34. Parotid gland
35. Sublingual gland
36. Submandibular gland
37. Lacrimal gland

N II. Optic nerve
N III. Oculomotor nerve
N VII. Facial nerve
N IX. Glossopharyngeal nerve
N X. Vagus nerve
Gg V. Semilunar ganglion
Gg VII. Geniculate ganglion
V_1. Ophthalmic nerve (V_1).
V_2. Maxillary nerve
V_3. Mandibular nerve
CB. Ciliary body
L_1. Lens at far vision
L_2. Lens at near vision

Cardiac Pain Fibers   Figure 30

Pain fibers from the heart (1, 2) have their cell bodies (or perikarya) in the dorsal root ganglia of thoracic spinal nerves T_1-T_5. In the drawing these are indicated by "1p" and "2p". Locate the termination of pain fibers "1" and "2" on the heart and trace and color them centrally. Note that fiber "1" ascends to the inferior cervical ganglion and turns down within the sympathetic trunk until it reaches one of the upper 4 or 5 thoracic ganglia. It then leaves the sympathetic trunk via a white ramus communicans, enters the spinal nerve at that level, proceeds centrally by way of the dorsal root, and finally enters the spinal cord where it synapses upon a second order neuron "11".

Trace pain fiber "2" as it leaves the heart and ascends to the middle cervical ganglion where, like fiber "1", it passes through with no interruption and descends to the level of thoracic nerve T_2. It also leaves the sympathetic chain and enters the spinal cord by the dorsal root of the corresponding spinal nerve.

Each of the 3 cervical sympathetic ganglia in the neck give off cardiac branches to the heart. The cardiac branch from the superior cervical ganglion is believed to carry no pain fibers, only efferents. The efferent sympathetic fibers arise from the 3 cervical sympathetic ganglia plus the upper 5 thoracic sympathetic ganglia (the first thoracic ganglion is often missing having fused with the inferior cervical ganglion to form the stellate ganglion). The sympathetic fibers from these ganglia converge on the heart where they form the cardiac plexus (the smaller top figure shows an exaggerated cardiac plexus; in reality it is found at the base, i.e., back, of the heart). The outflow (efferents) of the sympathetic nervous system occurs only at thoracic nerves T_1-L_2.

Locate the perikaryon (3p) of preganglionic sympathetic neuron "3" and note it lies in the intermediolateral column of the spinal cord gray matter. All preganglionic sympathetic perikarya are located in the intermediolateral column. Follow the axon of "3" out the ventral root, through the white ramus communicans into the sympathetic chain where it synapses upon postganglionic neuron "7" which sends its axon (cut) to the cardiac plexus. Fiber "3" continues to ascend up into the neck where it makes synaptic contact upon postganglionic neurons "6", "5" and "4" in the inferior, middle, and superior cervical ganglia respectively.

Neuron "8" in thoracic sympathetic ganglion T_2 also sends its axon to the cardiac plexus.

The postganglionic sympathetic fibers release norepinephrine (noradrenalin) which increases the force and rate of the heart beat and dilates the coronary arteries. Preganglionic sympathetic neuron "9p" lies in level T_6 of the spinal cord. Its axon will become a thoracic splanchnic nerve. The greater splanchnic receives fibers from thoracic ganglia T_5-T_7, most of which are preganglionic. The postganglionic sympathetic cell bodies (10) are located in ganglia such as the celiac which are also called prevertebral ganglia.

The preganglionic parasympathetic fibers travel within the vagus nerve (not shown). The postganglionic parasympathetic fibers (not shown) are within the walls of the heart. Afferent fibers from the heart that mediate cardiac reflexes travel with the vagus. Unlike the sympathetic pain fibers, these vagal cardiac afferents function at the subconscious level. Removal of the upper 4 or 5 thoracic sympathetic ganglia bilaterally should interrupt the path of the cardiac pain fibers and abolish pain from the heart (plus effective sympathetic influence on the heart).

Central Visual Pathways (N.V (N.II) Figure 31

If you were to encounter a penguin holding a sign "penguin" and standing in front of another sign that said "up" (top figure), the penguin and signs would constitute your visual field. That is, your visual field is what you see. By staring at the penguin's perplexed expression you would bring his face to be projected upon the central portion of your retina, the macula lutea or "yellow spot". The macular portion of your visual field is indicated by the lighter area that encompasses the penguin's face.

The biconvex lens in each eye will focus the image of the penguin upside-down on the retina on the back of each eye. The center of your visual field, in this case the penguin's face, will fall upon the macula lutea. Hence, the retina fields in each eye will each be an upside-down penguin.

Ganglion cells in the retina give rise to the axons that comprise the optic nerve (N II) and convey visual impulses centrally to a nuclear mass in the back of the thalamus, the lateral geniculate body.

Each eye is drawn so that only its bottom half is shown. The two inverted penguins are the retinal fields. The enlarged inserts behind each eye show that the penguin's eyes and beak fall in the macular field, one eye and half of the beak in each of the four quadrants of the macula.

Note that the "optic nerve" only extends to the chiasma; beyond the chiasma it is called "optic tract". (Actually, the retina is embryologically derived from the diencephalon and, hence, is a part of the brain; thus the whole optic nerve-tract should be called "tract". Furthermore, the optic "nerve" is the only nerve to be ensheathed by the three meninges and surrounded by a subarachnoid space with cerebrospinal fluid.)

Four ganglion cells (1, 2, 3, 4) are shown. Note that the two ganglion cells that arise from the lateral or temporal retinal fields (1, 4) have axons that project to the lateral bodies on the same side. That is, they do not cross but remain ipsilateral. The fibers that arise from ganglion cells in the medial or nasal retinal fields (2, 3) cross at the chiasma and project to the opposite or contralateral lateral geniculate body.

Neurons within the lateral geniculate body give rise to axons that comprise the optic radiation or geniculocalcarine tract which is a large curved bundle of fibers arising in the lateral geniculate body and ending mainly in the medial surface of the occipital lobe of the cerebral hemispheres.

The lateral geniculate body is divided into six layers or laminae of which layers 2, 3, 5 receive uncrossed fibers from the lateral half of the ipsilateral retina and layers 1, 4, 6 receive fibers from the medial half of the contralateral retina. Thus ganglion cells "1" and "3" which both lie in the left half of each retina project to the left lateral geniculate body and ganglion cells "2" and "4" that lie in the right half of each retina project to the right lateral geniculate body.

Because of this arrangement, half the penguin will be projected to each lateral geniculate body. The macular projection occupies the wedge-shaped posterior central region of the lateral geniculate body and the peripheral retinal fields occupy the remaining medial and lateral parts.

The drawing shows an image of each half of the penguin occurring twice in each lateral geniculate body. The macular projection which is half of the penguin's beak and one eye is disproportionately large. It occupies layers 5 and 6 with corresponding parts lying side by side but in separate layers. The rest of the penguin is shown falling in layers 3 and 4. It is probable that each half of the retinal field is represented not twice, but four or even six times, once in each layer of the lateral geniculate body.

In the visual cortex the macular representation is exceedingly large and located most posteriorly. In the drawing each cortical macula is represented as a greatly enlarged upside-down eye and half a beak. The rest of the penguin which falls within the peripheral fields is located more anterior and is quite small, especially when compared with the large size of the peripheral visual field and peripheral retinal field. In terms of what you look at, the world about you is projected onto your visual cortex upside-down with the right half of the visual field ending in the left half of the brain, and the left visual field ending in the right half of the brain. Note that the sign "up" is now down and entirely within the right lateral geniculate body and right visual cortex since in the visual field it lies in the left half.

Directions: Use one color for ganglion cells "1", "3" and the left optic radiation, and another for ganglion cells "2", "4" and the right optic radiation.

Central Connections of the Optic Nerve (N.II) Figure 32

The two spheres at the top represent the posterior aspect of the two retinas each of which is divided into four peripheral non-macular quadrants (a to h) and four central macular quadrants (s to z). The neurons shown in the retinas are ganglion cells and their axons are the optic nerve fibers (N II).

One ganglion cell lies in each of the four macular quadrants as well as in the four non-macular quadrants. The ganglion cells and the neurons of the lateral geniculate body and their respective fibers portray in a very crude and simplified manner the orderly projection of the retina to visual cortex (the retinotopic projection).

In the drawing each optic nerve contains eight fibers, whereas in the human optic nerve there are an estimated one million fibers.

Locate ganglion cell "a" in the lower left non-macular retina in the right eye and trace its fiber centrally. Note that it crosses to the opposite side at the optic chiasma and ends in the opposite lateral geniculate body where it synapses upon another "a" neuron in layer four.

Each lateral geniculate body consists of six layers with layers 1, 4, and 6 receiving fibers from the opposite eye, and layers 2, 3, and 5 receiving fibers from the homolateral eye. Other fibers from the medial (or nasal) half of the right eye will also end in layer 1 or layer 6, as well as layer 4 of the left lateral geniculate body.

Note the bend in the lateral geniculate body ("geniculate" in Latin means "bent", genu being "knee"). Its concavity is directed anteromedial and somewhat inferior.

Fibers from the macular region end in the central region of the lateral geniculate body, with those from the upper macula ending in the posteromedial half of the central region (v, u, w, x) and those from the lower half of each macula ending in the anterolateral half of the central region (s, t, y, z).

Locate ganglion cell "b" in the left retina. It is in the homologous position to cell "a"; that is, in the lower left non-macular retinal field of the left eye. Retinal field "a" will receive almost the identical visual image as does retinal field "b". In other words, they each see the same thing, although at a slightly different angle. This difference in visual images between the two eyes is essential to three-dimensional perception.

Note that the fiber arising from ganglion cell "b", unlike that of "a", remains uncrossed and ends in the homolateral lateral geniculate body in layer 3 close to the termination of fiber "a" in layer 4.

Trace the fibers of ganglion cells "c" and "d". Note that they end in layers 4 and 3 respectively of the left lateral geniculate body (labelled l sup n mac in drawing, left superior non-macula).

Trace the paths of the other homologous pairs from non-macular ganglion cells, "e" and "f", "g" and "h". Note that these all arise from the right half of each retina and end in the right lateral geniculate body.

Locate ganglion cell "s" in the right eye and trace it to its ending in layer 6 of

51

the <u>left</u> lateral geniculate body.

Find the homologous ganglion cell "t" in the left eye and follow its fiber to layer 5 of the left lateral geniculate body. Note that both "s" and "t" end in the anterolateral half of the center of the left lateral geniculate body (l inf mac, left inferior macula).

Locate ganglion cells "u" and "v", both in the upper left maculae, and note that they terminate in the posteromedial part of the macular region of the left lateral geniculate body.

Trace out the other homologous macular ganglion cells, "w" and "x", "y" and "z" to their terminations in the macular region in the right lateral geniculate body.

Now trace the course of the fibers that comprise the optic radiation (or geniculo-calcarine tract; <u>calcar</u> is Latin for spur; the <u>calcarine</u> sulcus bears a fanciful likeness to a spur).

The fibers of the optic radiation arise from cell bodies in the lateral geniculate body and terminate in the primary visual cortex most of which lies on the medial surface of the occipital lobe. This area is designated area "17" by Brodmann. It is also called the "striate cortex" because of the prominent line visible with the naked eye in this cortical region. The calcarine sulcus divides the visual cortex into upper and lower halves.

The upper retina projects to the upper visual cortex, and the lower retina to the lower visual cortex. The macula which is quite small in the eye has large cortical representation at the posterior pole of the brain. The left half of each retina projects to the left occipital cortex, and the right half to the right occipital cortex.

Note that fibers which carry impulses from the lower non-macular portions of the retina (a, b, g, h) make a forward loop before turning posterior to the visual cortex. These fibers actually travel ventrally and make a bend within the temporal lobe in forming this loop.

<u>Directions</u>: Use the same color for homologous pairs such as "a" and "b", and "s" and "t". Color the retinal quadrants, the regions of the lateral geniculate body, and the visual cortex the same color as the contained neuron. A suggested color scheme:

s, t	red	w, x	blue
a, b	light red	e, f	light blue
u, v	brown	y, z	green
e, d	light brown	g, h	light green

Cranial Nerves VI, IV, and III Figure 33

Six extrinsic ocular muscles move the eye and determine the direction of gaze. These are the superior rectus, lateral rectus, medial rectus, inferior rectus, superior oblique, and inferior oblique. A seventh muscle, the levator palpebrae, raises the eyelid.

These seven muscles are supplied by three cranial nerves, the abducent (VI), trochlear (IV), and oculomotor (III) nerves which contain 6,600, 3,400, and 24,000 fibers respectively.

The abducent nerve (N VI) arises from a cluster of motor somata that comprise the abducent nucleus (Nuc N VI) in the medial dorsal pons just under the floor of the IV ventricle. Neuron "1" is a motor neuron in the abducent nucleus. Its axon enters the orbit through the superior orbital fissure (not shown), passes through the anulus tendineus which is a ring of fibrous tissue at the back of the orbit from which the four recti muscles arise and supplies a single muscle, the lateral rectus (LR).

The action of the lateral rectus muscle is abduction of the eye, that is, turning the eye outwards or laterally. Damage to the abducent nerve or lateral rectus muscle will result in an inability of the eye to gaze laterally plus an inward turning of the eye when staring straight ahead due to the unopposed action of the antagonist medial rectus muscle.

The trochlear nerve (N IV), like the abducent nerve, also innervates only one muscle, the superior oblique. The trochlear nerve is unique among the cranial nerves in two respects: First, it is the only cranial nerve that completely decussates, thus innervating a structure on the opposite side; and, second, it is the only cranial nerve that emerges from the dorsal surface of the brain. The axon of neuron "2" in the left trochlear nucleus (Nuc N IV) crosses the midline and ends upon the right superior oblique muscle (SO). Like the abducent nerve and oculomotor nerve, the trochlear nerve passes through the superior orbital fissure, but unlike nerves VI and III which pass through the anulus tendineus, the trochlear nerve passes above the anulus.

Paralysis of the superior oblique will be most apparent when an attempt is made to direct the eye down and to the front as in walking down stairs.

The remaining five muscles are all innervated by the oculomotor nerve (N III) which arises from motor neurons in the oculomotor nucleus. This nerve also contains a contingent of preganglionic parasympathetic neurons in its visceral efferent nucleus of Edinger-Westphal. These preganglionic parasympathetic fibers leave the brain stem as part of the oculomotor nerve. (They were shown in figure 29.)

Both the oculomotor nucleus (Nuc N III) and trochlear nucleus (Nuc N IV) lie in the mesencephalon near the midline somewhat ventral to the cerebral aqueduct. The trochlear nucleus lies beneath the inferior colliculus and the oculomotor nucleus beneath the superior colliculus.

The oculomotor nucleus is partly paired and partly unpaired, the unpaired portion lies in the midsagittal plane and unites the two paired portions. While the majority of fibers within the oculomotor nerve remain uncrossed and supply muscles on the same side, surprisingly some fibers arise from somata in the opposite side of the nucleus.

Neurons "3" through "8" represent groups of motoneurons within the oculomotor nucleus each of which supplies one of five muscles. Their positions within the nucleus indicate the relative front-to-back localization of these neuronal groups.

Neuron "3" supplies the inferior rectus (IR). Neuron "4" supplies the medial rectus (MR) and neuron "5" innervates the inferior oblique (IO). Neuron "6" on the left side is exceptional since its axon crosses the midline to supply the right superior rectus. Hence, the superior rectus muscle is supplied by crossed fibers.

Neurons "7" and "8" both supply the levator palpebrae. Note that "7" is homolateral and "8" is on the opposite side. Thus the levator palpebrae receives both crossed and un-crossed fibers which may account for the fact that both eyelids are raised (and blinked) simultaneously.

A severance of the oculomotor nerve would result in paralysis of four extrinsic ocular muscles and the levator palpebrae. Hence, the eyelid could not be raised due to paralysis of the levator palpebrae. Drooping of the eyelid is called ptosis (literally "a falling" from the Greek, ptoma, to fall).

If the trochlear and abducent nerve were still functioning, the superior oblique muscle (N IV) would turn the eye down and the lateral rectus (N VI) would abduct it; thus with exclusive nerve III damage, the eye would be somewhat depressed and abducted.

The pupil constriction reflex in response to light would also be lost due to interruption of the parasympathetic fibers within the oculomotor nerve destined for the sphincter muscle in the iris.

Recent work has demonstrated the existence of internuclear cells within these three nuclei. The axons of these cells run from one nucleus to another. Neuron "9" is one such internuclear neuron; it lies in the left abducent nucleus and its axon ends in the right oculomotor nucleus upon a motor neuron which supplies the right medial rectus muscle. Internuclear neurons are probably responsible for coordinating conjugate eye movements; that is, movements in which both eyes move in the same direction. In order to turn both eyes to the left, the left lateral rectus and right medial rectus would work together. Conversely antagonist muscle must be inhibited; contraction of the right lateral rectus muscle must be accompanied by reciprocal inhibition of the right medial rectus. Thus internuclear neurons must have an inhibitory as well as excitatory role.

Directions: Using three different colors, color the fibers of abducent, trochlear, and oculomotor nerves each a different color.

The Cerebellum 1 General Plan Figure 34

A. <u>Top figure</u> is a highly simplified view of the dorsal aspect of the cerebellum consisting of three lobes, an anterior lobe (ant lobe), a much larger posterior lobe (post lobe), and a small flocculonodular lobe (fl nod fl).

The cerebellar cortex is divided into three longitudinal zones: A central <u>vermis</u> (Latin, worm); two <u>intermediate</u> (or paravermal) zones (int) on either side of the vermis, and the much larger <u>lateral</u> zones (lat) that cover the bulk of the cerebellar hemispheres.

The nodulus (nod) is part of both the vermis and flocculonodular lobe.

In terms of its functional connection, the cerebellum is divided into three parts: The <u>vestibulocerebellum</u> which is the flocculonodular lobe; the <u>spinocerebellum</u> which consists of most of the anterior lobe and a portion of the posterior lobe (indicated by the denser stippling around the words "leg" and "arm"); and the <u>pontocerebellum</u> which comprises the greater part of man's cerebellum, including the large hemispheres of the posterior lobe and the vermis between them.

"Arm" and "leg" in the spinocerebellar cortex are the sites of endings of spinocerebellar fibers that bring proprioceptive and tactile information from the arm and leg. Within both cortical areas that comprise the spinocerebellum, there is a complete somatotopic projection of the homolateral side of the body. The <u>pontocerebellum</u> consists of the remainder of the cerebellum (that is, everything that is not spinocerebellum or vestibulocerebellum).

The terms <u>archicerebellum</u>, <u>paleocerebellum</u>, and <u>neocerebellum</u> are often used for vestibulocerebellum, spinocerebellum and pontocerebellum, respectively.

<u>Directions</u>: Color the flocculonodular lobe, spinocerebellum and pontocerebellum each a different color.

B. The <u>bottom figure</u> shows a transverse section through the cerebellum and pons at the level indicated by the two arrows in the upper figure. Neurons "1" through "4" are Purkinje neurons in the cerebellar cortex. There are an estimated fifteen million Purkinje neurons in the cerebellum.

Note that their axons project centrally and each ends in one of the four cerebellar nuclei. These nuclei are, from lateral to medial, the <u>dentate</u> (den), <u>emboliform</u> (emb), <u>globose</u> (glob), and <u>fastigial</u> (fast).

<u>Directions</u>: Color Purkinje neuron "1" in the lateral cerebellar cortex and trace its axon centrally to its ending upon neuron "5" in the dentate nucleus. Use the same color for neuron "5" and its axon which enters the brachium conjunctivum (bc).

All outflow from the cerebellar cortex is by Purkinje axons, almost all of which end in one of the cerebellar nuclei. <u>Locate</u> neuron "2" in the intermediate cerebellar cortex and trace its axon deep to the emboliform nucleus where it ends upon neuron "6" which also sends its axon into the brachium conjunctivum. <u>Color</u> neurons "2" and "6" the same color.

Neuron "3" directs its axon to neuron "7" which lies in the globose nucleus and, like "6", also sends its axon into the brachium conjunctivum. <u>Color</u> neurons "3" and "7".

Most of the outflow of the cerebellum itself is by way of fibers in the brachium conjunctivum which arise from cell bodies in the dentate, emboliform, and globose nuclei.

Locate Purkinje cell "4" in the vermal part of the cortex. It projects to the fastigial nucleus whose neurons "8" and "9" project to the vestibular nuclei (vest). Note that "8" projects across the midline to the opposite vestibular nuclei whereas "9" projects to the vestibular nuclei on the same side.

Color neurons "8" and "9". Neurons "10", "11", and "12" are Purkinje cells in the flocculonodular lobe (fl nod fl). Purkinje cell "10" projects directly--that is, not by way of a cerebellar nucleus--to the vestibular nuclei. Purkinje cell "11" projects centrally to the fastigial nucleus, and Purkinje neuron "12" in the flocculus projects directly to the vestibular nuclei. Color flocculonodular neurons "10", "11", and "12". Neuron "13" is a pontine nucleus neuron whose axon crosses the midline and ends in the cerebellar cortex. Purkinje neurons "1", "2", "3", "4", and "11" give rise to cerebellar corticonuclear fibers. In addition to giving off corticonuclear fibers, each area of the cerebellar cortex also receives nucleocortical fibers from the same nucleus to which it projects.

C. This drawing depicts the three left cerebellar peduncles; the superior cerebellar peduncle (sup), the middle cerebellar peduncle (mid), and the inferior cerebellar peduncle (inf) drawn against the right half of the brain stem and the right half of the cerebellum (CBL). The three cerebellar peduncles each has another name, the superior peduncle is also called the brachium conjunctivum (Latin, "arm that is joined"), since the two peduncles decussate on their course to the red nucleus; the middle peduncle is also called the brachium pontis (Latin, "arm of the pons"); and the inferior peduncle is also called the restiform body (Latin, "rope-like"). Note the cap-like dentate nucleus sitting upon the superior peduncle.

The superior peduncle contains mostly efferent fibers that arise from cell bodies in the dentate, emboliform, and globose nucleus. The superior peduncle in man carries about 800,000 fibers and the dentate nucleus contains about 284,000 neurons. The middle peduncle is by far the largest consisting of about twenty million afferent fibers that arise from the contralateral pontine nuclei (pontocerebellar fibers). The inferior peduncle contains several incoming tracts with the largest being those from the opposite inferior olive (olivocerebellar fibers). Another large tract in the inferior peduncle is the dorsal spinocerebellar tract.

The Cerebellum 2 Cortical Neurons Figure 35

A. Sagittal view of the cerebellum and brain stem. Note how most of the cerebellar cortex is hidden from view by folds and subfolds. Beneath the outer cortex lies the white matter consisting of myelinated fibers running to and from the cortex. The encircled area is further enlarged in figure B. Key: 1, nodulus; 2, fourth ventricle; 3, pons; 4, medulla; 5, pineal gland; 6, medulla; 7, cerebral aqueduct.

B. Simplified view of the cerebellar cortex showing the three cortical layers: The outer molecular (mol), the intermediate Purkinje cell layer (Pur), and the inner granule layer (gr). Note the arrangement of the Purkinje cell dendrites (lc den) which lie in a plane that is transverse to the longitudinal axis of the folds of the cortex. These folds are called folia (Latin, folium means leaf). p, Purkinje cell somata.

C. Diagrammatic view of the cerebellar cortex. Afferents reach the cortex by only two kinds of fibers, mossy fibers and climbing fibers. Locate mossy fiber "1" at the bottom of drawing. Note that it gives off a collateral to neuron "13" which is a cerebellar nuclear neuron (such as those in the dentate nucleus or fastigial nucleus). Follow it as it ascends.

Note that it divides into several terminal branches, each of which makes synaptic contact with the dendrite of a granule neuron (2). Actually each such synaptic contact always has a third element which is the Golgi cell axon (shown enlarged in figure D).

The axon (3) of granule cell "2" extends up to the molecular layer where it divides like a "T" and forms parallel fiber (4). Parallel fibers run parallel to the long axis of the folium. In its course within the folium, a single parallel will make synaptic contact with the dendrites of Golgi cells (5), Purkinje cells (6), stellate cells (10), and basket cells (14).

The two branches of one parallel fiber probably exceed 3 mm (3,000 um) and one parallel fiber may act upon about 450 Purkinje neurons. A single Purkinje neuron may have as many as 400,000 parallel fibers passing through its dendritic tree, and hence be affected by a tremendous number of granule cells.

Note that the dendrites of the Purkinje cells lie in a plane that is transverse to the long axis of the folium. Note that the axon of Golgi cell "7" divides into a number of branches, each of which makes synaptic contact with a mossy fiber termination. Along with the granule cell dendrite, the mossy fiber ending and that of the Golgi cell form the cerebellar glomerulus (shown enlarged in figure D). The Golgi cell is an inhibitory neuron and tends to inhibit the firing of the mossy fiber so that the latter can no longer excite the granule cell and its parallel fiber. Hence, the Golgi cell acts as a negative feedback on mossy fiber activity.

Locate parallel fiber "9" near the top and note that it contacts the dendrites of stellate cell "10", which in turn makes synaptic contact witht the dendrites (11) of another Purkinje cell. The stellate cell is also inhibitory and tends to inhibit the Purkinje cell by inhibiting its dendrites.

Return to the bottom of the page and locate climbing fiber "12". Of the two types of fibers, mossy and climbing, that bring impulses into the cerebellar cortex, mossy fibers tend to branch extensively so that one fiber may send branches to more than one folium. Climbing fibers, on the other hand, are much more limited in their distribution and end within a rather restricted area. Climbing fibers arise almost entirely from the opposite inferior olive, whereas mossy fibers arise from a number of diverse sources such as the pontine nuclei and spinocerebellar tracts.

Follow the course of climbing fiber "12". Note that it also gives off a collateral to nuclear neuron "13". Note that it gives off additional branches to basket cell "14" and stellate cell "15". The main branch of each climbing fiber is directed to a Purkinje cell upon whose dendrites the climbing fiber "climbs" (17) somewhat like ivy on a tree. (Each climbing fiber may go to more than one Purkinje cell.)

Locate basket cell "14" on the lower right. Follow its axon which runs in the transverse plane and forms "baskets" (18) about the perikarya of several Purkinje cells. "Baskets" "18" are empty and indicate the site of two Purkinje cell perikarya. Purkinje cell "16" has an enveloping net of basket terminals. The basket cell, like the Golgi and stellate cell, is inhibitory and exerts an inhibitory effect on the Purkinje cell perikaryon. The Purkinje cell is also inhibitory. Its axon, which ends in one of the four nuclei, is the only efferent fiber from the cerebellar cortex. pc, Purkinje perikarya; gr, granule cell.

D. Cerebellar glomerulus. mf, mossy fiber; gr den, granule cell dendrite; Go ax, Golgi axon terminal.

Directions: Color each of the indicated components and all their processes a different color.

The Cerebellum 3 Afferents Dorsal View Figure 36

The drawing depicts the dorsal aspect of the cerebellum with the pontine nuclei (PN) and inferior olive (Inf O), both partially dissected, on the right. On the left are the four cerebellar nuclei, the dentate (D), emboliform (E), globose (G), and fastigial (F).

The perikarya of neuron "1" and "2" (bottom) lie in the nucleus dorsalis or Clark's column. Their axons turn laterally, ascend as the dorsal spinocerebellar tract (DSCT), enter the cerebellum via the restiform body, and end in one of the two "spinal" areas of the cerebellar cortex. Fiber "1" ends in the anterior "spinal" region which includes most of the anterior lobe, and fiber "2" ends in the posterior "spinal" region which includes the pyramis lobule of the vermis and the adjoining gracile lobule. (Actually, each "spinal" region receives a complete projection of the ipsilateral half of the body, but for simplicity's sake, each spinocerebellar tract is drawn with two neurons, one (1) going to the anterior "spinal" region and the other (2) to the posterior "spinal" region.) The dorsal spinocerebellar tract conveys exteroception and proprioception from muscle spindles, tendon organs, pressure receptors, and touch receptors in the leg and lower trunk.

The arm and upper trunk equivalent of the dorsal spinocerebellar tract is the cuneocerebellar tract. Its cells of origin (7, 8) lie in the external cuneate nucleus (ECN) which is the cervical equivalent of the nucleus dorsalis. The cuneocerebellar tract mediates proprioception and exteroception from the arm and upper trunk. Its fibers enter the cerebellum via the restiform body and end in the "arm" regions of the two "spinal" areas.

Neurons "5" and "6" give rise to the ventral spinocerebellar tract (VSCT) whose axons cross the midline and ascend on the opposite side. Hence, the ventral spinocerebellar tract is a crossed tract. Its fibers enter the cerebellum through the brachium conjunctivum and end in either of the two "leg" regions. The ventral spinocerebellar tract carries proprioception from the leg and lower trunk.

Recently, a fourth tract, the rostral spinocerebellar tract (RSCT), has been discovered. This is the arm equivalent of the ventral spinocerebellar tract and carries proprioception from the arm and upper trunk. However, unlike the ventral spinocerebellar tract which is largely a crossed tract, the rostral spinocerebellar tract (3, 4) is an uncrossed tract and ends in the ipsilateral "arm" regions.

All four spinocerebellar tracts end as mossy fibers in the cerebellar cortex.

Neurons "9" and "10" in the inferior olivary nucleus and neuron "11" in the medial accessory olivary nucleus (Med acc O) project to the opposite cerebellar cortex. Particular parts of the inferior olive each project to longitudinal strips or zones on the opposite cortex (indicated by arrows). Olivocerebellar fibers end as climbing fibers.

The largest afferent source to the cerebellum comes from the pontine nuclei. Pontocerebellar fibers cross the midline, group into the brachium pontis and project to the entire cerebellar cortex with the exception of the nodulus. Pontocerebellar fibers become mossy fibers and end in a pattern in which particular parts of the pontine nuclei each end in a somewhat circular zone suggested by "PC" around the termination of fiber "12". Neurons "12", "13", and "14" are pontine nuclei neurons and their axons are pontocerebellar fibers.

Directions: Color the different afferents to the cerebellum as indicated.

The Cerebellum 4 Afferents Lateral View Figure 37

This drawing of the cerebellar afferents corresponds closely to the preceding figure. Only the right side of the cerebellum and brain stem is shown. The "vacant" left side shows only the red nucleus (RN), brachium conjunctivum (BC), dentate nucleus (D), and cerebellar afferent fibers.

The olivocerebellar fibers (9, 10) and pontocerebellar fibers (11-15) arise from the right side and extend quite far laterally into the left part of the cerebellum (that is, towards the viewer). The spinocerebellar fibers end in the two spinal areas which are closer to the midline.

Why there are two spinal areas is not known; nor is it understood if one spinal area has a different function than the other.

An apparent paradox seems to apply to the laterality of the four spinocerebellar tracts: Three are homolateral and one is contralateral, the exception being the ventral spinocerebellar tract.

Of the four spinocerebellar tracts, most is known about the dorsal and ventral spinocerebellar tracts which were first described about one hundred years ago; the cuneocerebellar and rostral spinocerebellar tracts are rather recent discoveries. The dorsal and ventral spinocerebellar tracts differ in several respects other than their laterality: Dorsal spinocerebellar fibers are large in diameter, whereas ventral spinocerebellar fibers are thinner; dorsal spinocerebellar fibers carry information from a single muscle or small area of skin, whereas ventral spinocerebellar fibers carry information from a wide receptive field such as many muscle spindles (all the muscle spindles in the hindlimb may stimulate a single ventral spinocerebellar fiber); and finally, the fibers within the dorsal spinocerebellar tract end in a small area of the cerebellar cortex, whereas ventral spinocerebellar fibers each divide repeatedly, thus each supplies a much larger area of cerebellar cortex. Inf O, inferior olive; PN, pontine nuclei; T, tonsil of cerebellum.

Directions: Color the cerebellar fibers as indicated.

The Cerebellum 5 General Circuitry Figure 38

Neuron "1" (top) lies in motor area 4 in the precentral gyrus. Its axon is a corticopontine fiber that extends from the cerebral cortex to the pons where it synapses upon pontine nucleus neuron "2". The axon of "2" crosses the midline and projects to the opposite cerebellar cortex where it becomes a mossy fiber and synapses with granule cell "3".

The axon of granule cell "3" divides like a "T" into a parallel fiber (pf) that makes synaptic contact with the dendritic tree of Purkinje neurons "4" and "5". Purkinje neurons "4", "5", and "6" project their axons centrally to the dentate nucleus where they connect with dentate nucleus neurons "7", "8" and "9". Neuron "7" gives off a nucleocortical collateral fiber (ncf) that extends back to the cerebellar cortex.

The main axon of neuron "7" groups together with "8" and "9" as brachium conjunctivum fibers (BC), crosses the midline, and enters the opposite red nucleus within which fiber "7" gives off another collateral to neuron "12". The main fiber of "7", with that of "8", continue, rostrally to the ventralis lateralis thalamic nucleus (VL) where they end upon neurons "10" and "11" which project up to the motor cortex.

Thus, the loop of neurons just described, consisting of neurons in the cerebral cortex, pontine nuclei, cerebellar cortex, dentate nucleus, and ventralis lateralis thalamic nucleus, forms a "feed-back" mechanism in which motor commands issuing from the cerebral cortex are sent to the cerebellum where they are compared with all on-going muscle activity. The cerebellum appears to correct and modify these muscle commands via its projection back to the motor cortex.

Another "loop" is formed in part by neuron "12" in the red nucleus which receives a collateral of fiber "7" from the dentate nucleus. Neuron "12" projects caudally to the ipsilateral inferior olive (Inf O) as a rubro-olivary fiber. It synapses in the inferior olive upon neuron "13" which projects its axon, an olivocerebellar fiber, to the opposite cerebellar cortex where it ends as a climbing fiber (cf) upon Purkinje cell "6". The "loop" is completed by "6" projecting to "7" which projects back to "12". Purkinje neuron "14" (left center) projects to neuron "15" in the emboliform nucleus (E) which sends its axon via the brachium conjunctivum to the opposite red nucleus where it synapses upon neuron "25". Neuron "25" gives rise to the rubrospinal fiber which crosses the midline and descends in the lateral funiculus of the spinal cord (RST). Purkinje neuron "16" in the posterior vermis projects to the fastigial nucleus (F) where it contacts fastigial neuron "17". Note that fastigial neurons "17" and "18", which lie in the posterior part of the fastigial nucleus, both send their axon across the midline, through the opposite fastigial nucleus to the medial (M) and inferior (I) vestibular nuclei. Purkinje neuron "19" projects to fastigial neuron "20" which, together with fastigial neuron "21", projects to the ipsilateral vestibular nuclei, neuron "20" to the superior vestibular nucleus (S), and "21" to the medial vestibular nucleus (M). Purkinje neurons "22" and "23", both in the vermis of the anterior lobe, project to the lateral vestibular nucleus (L).

Note that "22" which is anterior ("leg") projects to the posterior part of the lateral vestibular nucleus, whereas posterior "23" ("arm") projects to the anterior part. Neuron "24" in the flocculus projects directly to the superior vestibular nucleus.

G: globose nucleus.

Directions: Color each group of neurons one color: 1-11; 12-13; 14, 15, 25; 16-24.

Symptoms of Cerebellar Disease Figure 39

Cerebellar dysfunction is worse if the superior cerebellar peduncle or dentate nucleus is damaged. If either of these is affected, the following symptoms are likely to occur.

A decrease in muscle tone or hypotonia. This can be detected by the examiner as a flabbiness when manually examined and as a marked lessening of resistance to passive movement. These deficits will be on the same side as the lesion, as is the case for all cerebellar symptoms.

The other symptoms of cerebellar disease are disturbances of rate, range, and force of movement (Holmes). These include dysmetria in which the hand overshoots its mark (past pointing) (figure A).

Rapid alternating movements, such as pronation and supination of the hand, are difficult to perform (adiadochokinesia) because of irregular muscle force and speed.

The conjugate gaze is jerky rather than smooth. The eyes overshoot their mark and oscillate before fixing correctly on an object. Nystagmus may or may not be present (figure C).

If the examiner's hand prevents the patient's limb from moving and the examiner suddenly removes his hand, the patient will be unable to check his limb movement and it will fly out. He may even hit himself in the face. This is the loss of the check reflex (Holmes' rebound phenomenon) (figure B).

The gait is wide-based, unsteady, and irregular. The patient may lurch from side to side (figure F).

Muscular incoordination (ataxia, Greek a without; taxis order) may be so severe that the patient cannot stand without assistance (figure F).

Movement does not proceed smoothly; rather each movement is broken up into a disjointed series of its constituent parts (decomposition of movement).

An "intention" tremor or "action" tremor is another cardinal symptom of cerebellar dysfunction. It is not a tremor in the classical sense as much as a swaying of the hand as it approaches the target. This is due, in part, to defective postural stabilization at the shoulder and elbow and, in part, due to the patient's attempts to correct the abnormal movement. The intention tremor occurs only during movement and becomes worse at the end of the movement (figure E).

Cerebellar dysarthria is a speech dysfunction in which the main feature is the slowness of speech. Words are not pronounced correctly and the syllables of individual words are unnaturally separated (scanning). There is also a lack of coordination between breathing and speech. Some words may be uttered with greater force than normal (explosive speech) and others may not have enough breath to be heard.

If the damage affects only the cerebellar cortex and the superior cerebellar peduncle and dentate nucleus are left intact, the disturbances of movement tend to be transitory and disappear with time.

Figures 39 A, B, C, D, F drawn by Janice Lalikos.

The Pyramidal Tract Figures 40 and 41

Neuron "1" is a cell of origin of pyramidal or corticospinal fiber in the "leg" region of the precentral gyrus (motor area 4). Trace its fiber caudally to its termination. Note how the three pyramidal fibers "1", "2", and "3" are "squeezed together" in the posterior limb of the internal capsule (asterisk *, on left side) between the thalamus (thal) and medial part of the globus pallidus (gp 1). Note also how the pyramidal fibers occupy the intermediate third of the cerebral peduncle in the midbrain. In the pons the pyramidal fibers are broken up into longitudinal bundles that regroup into the compact pyramids (pyr) in the medulla.

In the lower medulla between 75-90% of the fibers decussate (pyr decuss) and continue down the spinal cord as the lateral corticospinal tract. The remaining uncrossed fibers become the ventral corticospinal tract. Fiber "1" carrying leg "motor commands" continues caudally to the sacral level where it synapses upon internuncial (interneuron) "4" which in turn contacts alpha motor neuron "5" which sends its axon to muscle fibers in a muscle of the leg.

Corticospinal neuron "2" in the "hand" region of the motor cortex is exceptional because its axon ends directly upon alpha motor neuron "6" in the cervical spinal cord instead of upon an internuncial. Neuron "6" is an alpha motor neuron (or lower motor neuron) whose axon innervates muscle fibers in a muscle of the hand.

Neuron "3" is a corticobulbar neuron**. Its fiber extends from the "face" region of the "motor" cortex to the opposite facial nucleus where it synapses upon facial motor neuron "6" whose axon innervates muscle fibers in a muscle of facial expression such as the orbicularis oris.

By definition, the pyramidal or corticospinal tract is all those fibers that pass through the pyramids in the medulla. The pyramidal tract begins with cells that lie in the fifth layer of the cerebral cortex and ends in the spinal cord where most of its fibers synapse upon interneurons which influence both alpha and gamma motor neurons.

The pyramidal tract is only one of several descending motor tracts that carry "motor commands" to the lower motoneurons; however, it is the only descending tract that travels from cerebral cortex to spinal cord with no synaptic interruption. It is found only in the higher mammals and reaches its greatest development in humans.

Each pyramidal tract contains about one million fibers of which about 90% are thin with a diameter of 1-4 μm. About 90% of all pyramidal fibers are myelinated and 10% unmyelinated. Some 30,000 pyramidal fibers are very large, with diameters of 11-22 um. These large fibers are believed to arise from the 30,000 giant Betz cells in motor area 4. These large fibers are the longest fibers, extending from the cortex to sacral levels of the cord. Neuron "1" could be one such large fiber. In their course to the spinal cord, pyramidal fibers give off many collaterals to structures such as the striatum, thalamus, red nucleus, pontine nuclei, and reticular formation.

Originally it was believed that pyramidal fibers arose only from the Betz cells in area 4. After more accurate counts of the number of Betz cells (30,000) and the total number of pyramidal fibers (about 1,000,000) became available, this view could no longer be held. It now appears that only about 25% of the pyramidal fibers have their origin in area 4.

One study found area 4 contributing only 22%, whereas area 3 had 48%, area 1 17%, area 3 7%, and area 5 which is also in the parietal cortex 6%. This would indicate almost 80% of the pyramidal tract arises from the parietal cortex.

It is now felt that in addition to carrying "motor commands" to alpha and gamma motor neurons, the pyramidal has an important sensory function: It most likely regulates the transmission of sensory information to the brain, particularly finger and hand proprioception during delicate movement such as writing or playing a musical instrument. Presumably, the "sensory" pyramidal tract component would "sharpen" or "zero in" on proprioceptive information from the hand and fingers, possibly augmenting these sensations, while at the same time suppressing proprioception from the rest of the body.

This suggestion is supported by the fact that a large proportion of pyramidal fibers arise from the somatosensory cortex (that is, the parietal cortex). Further evidence for this somatosensory-modulator role of the pyramidal tract is the site of termination of these pyramidal fibers that arise in the parietal cortex: They end in the dorsal horn of the spinal cord where they presumably could affect incoming proprioceptive and exteroceptive information.

Pyramidal fibers carrying "motor commands" end upon interneurons in the intermediate gray matter of the spinal cord. These interneurons (or internuncials) act upon both alpha lower or peripheral) motor neurons as well as gamma motor neurons. Thus, corticospinal neurons exert their influence upon the lower motor neurons by way of an interneuron so that two synapses are involved and pyramidal neuron-to-alpha motor neuron path is said to be disynaptic (two synapses).

There are in humans and the monkey, however, some pyramidal fibers that act monosynaptically; that is, they end directly upon the lower motor neuron. These monosynaptic fibers end in the region of finger alpha motor neurons and are presumed to supply finger muscles.

Interneurons, such as "6", are considerably more structurally complex than indicated in the drawing with a vast intricate array of dendrites. These interneurons probably play a crucial role in all movement. Even the simplest movement requires an incredible timing and coordination of all the motor neurons involved. The sequence of impulses to the prime movers must be exactly right so that, for example, the hand is moved to the desired position. The antagonist muscles must relax to just the right degree so that the hand does not overshoot its mark. The synergist muscles must stabilize and work with the prime movers and the gamma motor neurons must maintain the correct tautness of the muscle spindle fibers so that the muscle spindles remain sensitive even as the muscle shortens.

The interneuron probably plays a key role in determining the exquisitely complex discharge pattern of the neurons involved.

The pyramidal tract is responsible for rapid, isolated, discrete movements of the fingers. Following pyramidotomy (cutting the pyramidal tract) in monkeys these movements are lost although the hand and fingers can still be used for simpler movements (for example, walking). The pyramidal tract's influence upon the gamma motor neuron helps maintain the normal muscle tone. Pyramidotomy results in hypotonia (lessened muscle tone) plus a diminution of the myotatic or stretch reflex. Damage to the pyramidal tract is not followed by muscle atrophy since the alpha motor neuron is still intact.

Put: putamen.

Directions: Trace the descending course of pyramidal fibers 1, 2, 3 using the same color. Use another color for alpha motor neurons 5, 6, 7 and a third one for interneuron 4.

**Corticobulbar fibers are usually discussed with pyramidal fibers even though, by definition, they are not pyramidal fibers; i.e., they do not traverse the medullary pyramids.

The Caudate Nucleus, Putamen, Globus Pallidus, and Amygdala Figure 42

The top figure A shows the relative position of these nuclear masses in a horizontally cut brain. The caudate (Latin, tailed) nucleus consists of a head (1), body (2), and a long tail (3) which curves posteriorly, then inferiorly, and finally anteriorly into the temporal lobe where it fuses with the amygdala (4) (Greek, almond). Note that the head of the caudate nucleus (1) is partially separated from the putamen (5) by the anterior limb of the internal capsule (10-11). Their anterior inferior portions are joined together. This is best seen in the lower two figures B and C which show these structures cut by two frontal planes.

Note that the amygdala (4) is continuous with the overlying putamen (5) as well as the tail of the caudate nucleus (3). This continuity of the amygdala with both the caudate and putamen has no apparent functional significance. Note that the caudate nucleus and putamen are joined by many thin cellular bridges that traverse the internal capsule (13).

The putamen (Latin, husk, shell) and inner globus pallidus (Latin, pale globe) constitute the lentiform (Latin, lens-shaped) nucleus. The lentiform nucleus, caudate nucleus and claustrum (Latin, cloak), which is a thin cellular mass lateral to the putamen, constitute the corpus striatum (Latin, striped body). The striated or striped appearance is due to the numerous bundles of myelinated axons that traverse these structures.

The caudate nucleus and putamen constitute the neostriatum (Greek, neos, new) or simply the striatum. The globus pallidus is the paleostriatum (Greek, paleos, old) or simply pallidum. The term basal ganglia (or, more correctly, basal nuclei) is often used synonymously with corpus striatum; that is, the caudate nucleus, lentiform (putamen and globus pallidus), and claustrum. The amygdala is usually not considered as one of the basal ganglia or as part of the corpus striatum.

Although not usually grouped with the basal ganglia, the subthalamic nucleus and substantia nigra (Latin, black substance) have abundant interconnections with the corpus striatum and some authors consider them as belonging to the basal ganglia.

The caudate nucleus and putamen contain an estimated 110 million closely-packed small neurons surrounding some 600,000 scattered large neurons. The globus pallidus, on the other hand, has considerably fewer cells, about 710,000, which are much larger and more widely dispersed.

```
     1.    Head of caudate nucleus
     2.    Body of caudate nucleus
     3.    Tail of caudate nucleus
     4.    Amygdala
     5.    Putamen
     6.    Globus pallidus (lateral part)
     7.    Globus pallidus (medial part)
     8.    Thalamus (left)
     9.    Outline of right thalamus
 10-11.    Anterior limb of internal capsule
    11.    Genu of internal capsule
 11-12.    Posterior limb of internal capsule
    13.    Cellular bridges between caudate nucleus and putamen
```

Directions: Use one color for the striatum (caudate nucleus, 1, 2, 3 and putamen 5.) Use a second for the globus pallidus (6, 7) and a third for the amygdala (4).

The Basal Nuclei. Interconnections 1   Figure 43

The main outflow from the corpus striatum is by two fiber tracts that arise in the medial segment of the globus pallidus and end in the thalamus. These are the fasciculus lenticularis and ansa lenticularis. Because they begin in the globus pallidus and leave it they are pallidofugal fibers (fugere, Latin, means to flee).

The fasciculus lenticularis begins in the dorsal part of the medial pallidal segment (neurons "1" and "2"), emerges from its dorsal medial surface, and travels across the internal capsule in little bundles which form the fasciculus lenticularis (fasc lent). The fibers continue somewhat posteriorly and converge in the space between the subthalamic nucleus (Subth nuc) inferiorly and the zona incerta (zona inc) superiorly where they lie in Forel's field H_2. They then proceed posteriorly and medially to the region in front of the red nucleus. This is the prerubral field H of Forel. Most of the fibers then turn superiorly and laterally to ascend between the zona incerta and the thalamus and in so doing become part of a larger bundle, the thalamic fasciculus. The thalamic fasciculus is the same as Forel's field H_1. In the drawing the thalamus is portrayed as hollow with only its border shown and part of this is cut away to show four fibers of the thalamic fasciculus, two of which (1 and 2) are derived from the fasciculus lenticularis.

Fibers that make up the ansa (Latin, handle) lenticularis begin with somata "3" and "4" in the ventral portion of the medial pallidal segment. They sweep rostrally and medially and make a loop around the internal capsule thus forming the ansa lenticularis (ansa lent). Continuing posteriorly they join fibers from the fasciculus lenticularis in the prerubral field H of Forel. Like the fasciculus lenticularis most of the fibers in the ansa ascend in the thalamic fasciculus and end in the thalamus. Fibers from the ansa end in the ventralis lateralis (VL) thalamic nucleus (neurons "5" in drawing); whereas those from the fasciculus lenticularis end in the ventralis anterior (VA) thalamic nucleus. This is not shown in the drawing and lies in a more rostral (anterior) plane than the section in the drawing. Neurons "5" in the VL nucleus project to the motor cortex (area 4).

Neurons "8" in the putamen send striatonigral fibers (from putamen to substantia nigra) which pass through the globus pallidus. Within the globus pallidus these striatonigral fibers give off collaterals to pallidal neurons "3" and "4". The main axons of "8" continue caudally and obliquely pass through the internal capsule to end upon neurons "13" in the substantia nigra. Caudate nucleus neuron "9" also gives rise to a striatonigral fiber which likewise sends a collateral to the pallidum where it ends upon pallidal neuron "1". The main fiber of "9" continues caudally, passes through the internal capsule and ends upon nigral neuron "13". Thus striatopallidal fibers are actually collaterals of striatonigral fibers.

Fiber "10" arises from premotor area 9 in the frontal lobe, and ends in the caudate nucleus. Fiber "11" arises from motor area 4, and "12" from somatosensory areas 3, 1, and 2, and both end in the putamen. Fibers "6" lie in the anterior limb of the internal capsule and fibers "7" in the posterior limb.

Thus the putamen and caudate nucleus receive impulses from the cerebral cortex and relay these to the globus pallidus which in turn sends them to the VA/VL thalamic nuclei which send them back to the cerebral cortex.

The Basal Nuclei. Interconnections 2 Figure 44

Neuron "1" in the head of the caudate nucleus sends its axon through the lateral portion of the globus pallidus (GP) where it gives off a collateral to a pallidal neuron. This striatofugal fiber continues caudally along the ventrolateral edge of the internal capsule. As the internal capsule becomes the cerebral peduncle it then turns dorsomedially through the peduncle to reach the pars reticularis (ventral part) of the substantia nigra where it ends. The caudate nucleus and putamen are the striatum, thus fiber "1" is a striatonigral fiber.

Recent work suggests that the striatonigral projection, or at least some axons in it, use GABA (gamma amino butyric acid) as their neurotransmitter. Additional work has shown that substance P, a neuropeptide of eleven amino acids, is also transported in striatonigral fibers. Thus fiber "1" could possibly be carrying GABA and thus be GABAergic, or it could be carrying substance P.

Fiber "2" arising from nigral neuron "2" is a nigrostriatal fiber. It arises from the dorsal pars compacta of the substantia nigra and ends upon a neuron in the head of the caudate nucleus. Neurons "3" and "5" are also nigrostriatal neurons; these, however, pass to the putamen. Nigrostriatal fibers such as fibers "2", "3", and "5" have been shown to use dopamine as their neurotransmitter and function as the major source of striatal dopamine; thus these fibers are dopaminergic. The degeneration of this nigrostriatal pathway is believed to be an important factor in the cause of Parkinson's disease.

Neuron "4" in the substantia nigra sends a nigrothalamic fiber to the ventral lateral thalamic nucleus (VL).

Only a few of the connections to and from the basal nuclei and related structures are shown in this drawing. There are numerous other projections such as a nigrotectal projection from substantia nigra to the superior colliculus. The substantia nigra has also been shown to project to such distant structures as the olfactory tubercle, amygdala, and even the frontal cerebral cortex.

Parkinson's disease is caused by degenerative changes in the substantia nigra. Pathological changes involve a loss of neurons and an interruption in the nigrostriatal dopaminergic pathway (see figure 45).

Abbreviations: DM, dorsomedialis nucleus of thalamus; GP, globus pallidus; Pul, pulvinar of thalamus; Subth nuc, subthalamic nucleus; VL, ventralis lateralis thalamic nucleus; VPL, ventralis posterolateralis thalamic nucleus; ZI, zona incerta.

Directions: Use one color for striatonigral neuron "1", another for nigrostriatal neurons "2", "3", and "5" and a third color for nigrothalamic neuron "4."

Parkinson's Disease Figure 45

Parkinson's disease (also called paralysis agitans) is the result of the loss of pigmented cells in the substantia nigra and other pigmented nuclei such as the locus ceruleus. These pigmented neurons release dopamine at their terminals in the striatum and their degeneration results in a depletion of dopamine in the caudate nucleus and putamen. The term "parkinsonism" refers to the group of symptoms that characterize this disease. These are essentially disorders of posture and movement and include an expressionless face, rigidity, poverty and slowness of movement, a resting tremor, stooped posture, and a shuffling gait.

The expressionless, masked-like face conveys no emotion and the somewhat widened palpebral fissures of the eyes impart a "staring" appearance. This is enhanced by a marked infrequency of blinking.

The rigidity is best detected by the examiner passively moving a limb. As the limb is moved a mild resistance can be felt. This resistance gives way in a series of small jerk-like movements each of which is momentarily arrested by the resisting muscles. The lengthening of the muscles occurs as a series of alternating short stretches and mild contractions. To the examiner, this feels somewhat like a cogwheel.

A poverty and slowness of all voluntary movement includes an absence of arm swing in walking, an infrequency of swallowing, a slowness of chewing, an absence of "adjusting movements" in both sitting and standing. As the disease progresses, it may take an hour to eat a meal. Handwriting becomes cramped and small (micrographia). The voice weakens until it is only a whisper.

Walking is reduced to a shuffle with the patient frequently losing his balance. To avoid falling, the patient "chases his center of gravity" and an intended walk turns into an ever-accelerating run (festinating gait, Latin, festinare, to hasten).

The characteristic tremor usually afflicts the hand and takes the form of the four-per-second "pill-rolling" tremor of the thumb and fingers. It is called a "resting tremor" because it is present when the hand is held motionless. However, it disappears during complete relaxation and is made worse by excitement. It is suppressed momentarily by voluntary movement.

"The tremor interferes surprisingly little with voluntary movement; it is not uncommon, for example, to see a patient who has been trembling violently raise a full glass of water to his lips and drain its contents without spilling a drop."*

The stooped posture is presumably caused by the hypertonus of the flexor muscles of the trunk and limbs.

Drawn by Joseph Kanasz

---

*R. Adams and M. Victor, Principles of Neurology. McGraw-Hill, N.Y., 1977, page 63.

Chorea   Figure 46

Chorea is the name for convulsive movements of a forced and rapid nature that suggest a grotesque dance. The word chorea means "dance" in Greek. Choreiform movements are apparently caused by the degeneration or the malfunction of neurons in the striatum with the caudate nucleus most frequently implicated.

The two most studied choreas are Huntington's chorea and Sydenham's chorea.

Huntington's chorea (also called Huntington's disease, chronic progressive chorea, adult chorea, chorea major) is a hereditary disease which appears between 35 and 45 years of age. It is characterized by choreiform movements and mental deterioration. The disease is transmitted from parent to offspring as a dominant factor and children of one parent with Huntington's chorea have a 50% chance of inheriting the disease.

The choreiform movements are sudden, purposeless and jerky with the trunk muscles and hip and shoulder muscles most often affected. Facial grimacing is common in the early stages of the disease (Figures on right).

"As the disease progresses, twisting and lordotic movements of the trunk (rotating the front of the pelvis downward) especially on walking associated with more intense arm and leg movements give the individual a dancing, prancing type of gait which is particularly characteristic of the disorder."*

The abnormal movements are made more severe by emotional situations and disappear during sleep.

The mental deterioration involves progressive memory impairment, failing intellect, apathy (loss of interest), and disregard for personal cleanliness.

Emotional abnormalities include irritability, impulsive behavior, depression, and even fits of violence. The mental deterioration and bizarre emotional behavior invariably result in commitment to a mental institution.

There is no cure for the disease which is progressive. It ends in death about 15 years from the time of its first symptoms. Suicide is not uncommon and children of Huntington's disease victims frequently forego having any children of their own since there is no way of determining if they themselves have inherited the disease until they are almost 50 years of age.

Sydenham's chorea (also called acute chorea, St. Vitus' dance, chorea minor, rheumatic chorea), when compared to Huntington's chorea, is a relatively benign disease of childhood. It is rarely fatal. Females are afflicted over twice as frequently as males.

The disease bears a close relationship to rheumatic fever and the choreiform movement and mental and emotional factors are most likely caused by the same streptococcic bacillus that causes rheumatic fever gaining access to the brain and somehow affecting neurons in the caudate nucleus and subthalamic nucleus.

The choreiform movements are quick, jerking, and flinging, and usually affect the arms more chan the legs or face. As is the case in Huntington's disease, the severity of the abnormal movements is greatly increased by emotional stimuli.

The child often displays mental and emotional changes such as irritability, apathy, and emotional instability. In severe cases, the patient may experience mental confusion, hallucinations, and delusions.

Drawn by Joseph Kanasz.

---

*Merritt, Hiram H. A Textbook of Neurology. Lea and Febiger, Philadelphia. 1979.

Athetosis and Hemiballism Figure 47

Athetosis (Greek, not fixed) is a peculiar type of dyskinesia (abnormal movement) resulting from degeneration in the caudate nucleus, putamen, and globus pallidus.

The movements are slow, twisting, writhing, and worm-like with the arms and hand most frequently involved.

A typical athetoid movement is shown in Figure 47 in which the forearm and hand alternate between flexion-supination and extension-pronation.

Hemiballism (or hemiballismus) is a violent flinging about of the arm or leg on the same side of the body. It may be confined to a single extremity such as the arm as shown in Figure 47. The term hemiballism means "half throwing" in Greek.

The shoulder and/or hip muscles undergo sudden and violent contraction which results in the limb(s) being thrown about aimlessly. The movements are more forceful and violent than are those of chorea and their persistence may lead to death from cardiac failure or exhaustion.

The causative lesion always involves the opposite subthalamic nucleus or its connections.

Drawn by Joseph Kanasz.

Stria Terminalis and Ventral Amygdalofugal Pathway Figure 48

The central figure A shows the left amygdala (also called amygdaloid complex) and left stria terminalis (stria term) superimposed upon the right half of the brain. The two figures on the right (B, C) are similarly oriented.

The stria terminalis is visible with the naked eye as it curves in an arc along the medial surface of the caudate nucleus in the lateral ventricle beginning at the amygdala and ending just above the anterior commissure (ac).

The two figures on the left (D, E) which show the right stria terminalis give a more accurate portrayal of its actual size. The lower left figure E is a frontal section of the right amygdala at plane X-X in the cental figure and plane Y-Y of the upper left figure D which depicts the right amygdala and related structures.

The stria terminalis is a bundle of mainly efferent fibers that arise mainly from the corticomedial portion of the amygdala and end mainly in the hypothalamus. Its fibers have been grouped into three categories: (1) commissural which cross via the anterior commissure to the opposite amygdala, (2) precommissural which pass in front of the anterior commissure, and (3) postcommissural which pass behind the anterior commissure.

In the large central figure A neuron "1" in the corticomedial amygdala represents the commissural component. Its fiber travels within the stria terminalis (stria term) to the anterior commissure (ac) which it enters. It then crosses to the opposite side, travels backwards in the opposite stria terminalis and finally ends in the cortical nucleus of the opposite amygdala.

The precommissural component (fiber 2) also arises from the corticomedial nuclear group (2), travels within the stria terminalis, and passes in front of the anterior commissure. Its fibers then pass backwards and downwards to end in the medial preoptic hypothalamic nucleus (4).

Postcommissural fibers (3) which may arise from either the lateral, basal or medial amygdaloid nuclei pass behind the anterior commissure and end in either the anterior hypothalamic nucleus (5) or the bed nucleus of the stria terminalis which is a collection of neuronal masses that accompany the stria terminalis in its curved course (figure C).

Figure B shows three hypothalamic nuclei that receive precommissural fibers. Neuron "6" projects to the medial preoptic hypothalamic nucleus (7). Neuron "8" projects to the anterior hypothalamic nucleus (9) and neuron "10" projects to the ventromedial hypothalamic nucleus (11). Figure C shows two postcommissural neurons "13" and "14". "13" projects to the bed nuclei of the stria terminalis (which remains posterior to the anterior commissure and lateral to the columns of the fornix, thus fibers such as "13" are postcommissural). Neuron "14" projects to the anterior hypothalamic nucleus. Neuron "12" in the cortical amygdaloid nucleus projects via the anterior commissure to the opposite cortical nucleus; thus "12" is a commissural neuron.

The ventral amygdalofugal pathway refers to a large assortment of fibers that leave the amygdala in a medial direction and then pass along the base of the brain to widely diverse sites. The two figures on the left (D, E) show three representative neurons of the ventral amygdalofugal pathway. Neuron "15" projects to the dorsomedial thalamic nucleus (16). Neuron "17" projects to the lateral preoptic nucleus (18) and neuron "19" which is cut off projects to the medial cerebral cortex in front of and below the genu of the corpus callosum.

These drawings depict only a few of the efferent connections of the amygdaloid complex. The amygdala projects caudally to the dorsal nucleus of X, the nucleus of the locus ceruleus, the reticular formation, the substantia nigra, and the periaqueductal gray. The amygdala also projects to the olfactory bulb and other areas of the cerebral cortex in addition to the medial frontal cortex mentioned above (fiber "19").

If the ventral amygdalofugal projection is completely interrupted defensive reactions (rage and aggression) can no longer be elicited from experimental animals. This suggests that the basolateral part of the amygdala plays an important role in defensive behavior. Electrical stimulation and ablation of the amygdaloid complex result in a wide variety of behavioral, visceral, somatic, and endocrine changes.

Abbreviations: ac, anterior commissure; gp, globus pallidus; op ch, optic chiasm; put, putamen; thal, thalamus.

Directions: Color the commissural fibers (1, 14), precommissural fibers (2, 6, 8, 10), postcommissural fibers (3, 13, 14), and ventral amygdalofugal fibers (15, 17, 19) each a different color.

Some Afferent Connections to the Amygdaloid Complex   Figure 49

The left amygdaloid complex is shown in lateral view at the bottom (A). It is cut at line "X" into two parts to reveal some of its component nuclei. The cortical nucleus (cort) and medial nucleus (med) which are small in humans probably have an olfactory-related function since they receive afferents from the olfactory bulb (1) and anterior ol-factory nucleus (2). The basal nuclear group (b) and lateral nucleus (lat), both of which are large in humans, probably function in a non-olfactory capacity. The larger drawing B shows the left amygdala and some afferent fibers superimposed on the right half of the brain. The amygdaloid complex receives afferents from widely diverse sites.

Directions: Color and trace neurons "1-11" which represent the following afferents of the amygdala. Fiber "1" arises from the olfactory bulb and ends in the corticomedial group of the amygdala. Fiber "2" arises from the anterior olfactory nucleus and also ends in the corticomedial group. Neuron "3" in the primary olfactory cortex projects to the basolateral group, as does neuron "4" in the anterior cingulate gyrus.

Neuron "5" in the medial prefrontal cortex reaches the basal, lateral, and central amygdaloid nuclei by a peculiar route: This fiber enters the stria terminalis which it leaves at the midthalamic level and descends through the internal capsule and globus pallidus on its way to the amygdala. Neuron "6" in the inferior temporal gyrus directs its fiber to the basolateral group. Neuron "7" in the dorsomedial thalamic nucleus directs its fiber to the basolateral amygdala. Neuron "8" in the ventromedial hypothalamic nucleus projects to the central nucleus as do both neuron "9" in the dorsal raphe nucleus and neuron "10" in the locus ceruleus. Neuron "11" in the lateral preoptic hypothalamic nucleus reaches most of the amygdaloid nuclei via the stria terminalis.

It has been known for many years that uncinate epileptic seizures often begin with an aura of an olfactory hallucination. The uncus (Latin, hook) lies on the medial temporal cortex directly over the amygdala. An epileptic seizure having its focus near or in this area would presumably trigger off neurons in the amygdala. The patient would actually "experience", or be convinced that he is perceiving, an odor even though there is no odor stimulating his nose. Usually the smell "experienced" is disagreeable such as burning rubber and sometimes it seems to originate from part of the body so that the patient will report that he "smells burning rubber in my stomach". Uncinate seizures may also give rise to emotional hallucinations in which the victim is subjected to overpowering feelings of dread or terror. Seizures of this type may have their focus in the basolateral nuclei, whereas the olfactory hallucinations probably involve the corticomedial nuclei or the nearby olfactory cortex. A number of findings suggest that the amygdala is involved with aggressive behavior among which is the now classical experiment of Kluver and Bucy who removed the temporal lobe of animals. One of their most striking findings was that following temporal lobe ablation, which included removal of the amygdala, normally aggressive animals became placid and tame.

The Hippocampal Formation   Figure 50

The smaller figure on the left (A) shows the left hippocampal (hippocampus, Greek, means sea horse) formation, left fornix (Latin, arch), and left mammillary (Latin, breast-like) body superimposed upon the right half of the brain.

Color the hippocampal formation, fornix, and mammillary body. Note that the hippocampal formation which is three-layered primitive cortex lies in the temporal lobe of the brain and is connected to the diencephalon, especially the mammillary bodies, by the arched fornix. The fornix contains about 1,200,000 fibers most of which arise in the hippocampal formation and end in a number of sites. Some fibers within the fornix, however, begin outside the hippocampus, travel in the opposite direction and end in the hippocampus. The llarger figure (B) shows some of the hippocampal neural connections. The two longitudinal components of the hippocampal formation which are the dentate (Latin, "toothed") gyrus and the hippocampus proper (also called Ammon's horn) are shown in highly schematic form with the subiculum which is actually part of the parahippocampal gyrus.

Color and trace the following: Neuron "1" in the entorhinal cortex projects to dentate gyrus neuron "2" which sends its axon to the hippocampus proper. Neuron "3" in the hippocampus proper projects to the subiculum. Neuron "4" in the subiculum gives rise to a fiber that enters the arched fornix. The axon of "4" passes in back of the anterior commissure (ac) and ends in the mammillary body which is part of the hypothalamus. Thus it is part of the postcommissural component of the fornix. Neuron "5" in Ammon's horn also contributes to the fornix and ends in neuron "8" in the septal nuclei. Neuron "6" in the parasubiculum projects by way of the fornix to the anterior thalamus (neuron "9"). Neuron "7" in the cingulate gyrus projects via the cingulum (Latin, "girdle") in a sweeping arch backwards and downwards to end in the subiculum.

For over one-hundred years it was believed that the great majority of fibers within the fornix, especially those that terminate in the mammillary bodies, arose from the pyramidal cells in the hippocampus proper. However, recent work that employed radioactive tracers has shown this not to be the case; rather, all fibers destined for the mammillary bodies arise from in the subiculum and the pyramidal cells of the hippocampus project largely to the septal nuclei which are located in front of the columns of the fornix.

The Hippocampus and Fornix Figure 51

Figure A and figure B are a frontal and medial view respectively of a dissected brain with the right hippocampal formation, fornix, and mammillary body exposed.

Color the hippocampal formation, fornix, and mammillary body in both figures A and B.

Central figure C shows both hippocampal formations, fornices, and mammillary bodies in frontal view. The left hippocampus (viewer's right) has been cut to reveal some of the internal structure. Locate neuron "1" in the subiculum, trace its axon as it enters first the alveus (Latin, a hollowed structure), then the fimbria (Latin, border) which is the first part of the fornix and lies upon the hippocampus proper, and finaly the crus (Latin, leg) of the fornix. The axon of subiculum neuron "1" ends in the mammillary body where it synapses upon neuron "2".

Return to the hippocampal formation and locate neuron "3" which is one of the pyramidal neurons in Ammon's horn. Trace the course of the axon of pyramidal neuron "3" to its termination upon neuron "4" in the septal nuclei.

Figure D shows a "disassembled" hippocampal formation. Notice that the fimbria is made up of myelinated axons which first pass on the ventricular surface of Ammon's horn as the alveus. Notice how both the hippocampus proper and dentate gyrus have the shapes of two interlocking "C's". Notice also that the hippocampus proper is characterized by the pyramidal cells and the dentate by granule cells.

The subiculum lies beneath (inferior to) the dentate and is continuous laterally with Ammon's horn and medial with the presubiculum which in turn gives way to the parasubiculum. The subiculum, presubiculum, and parasubiculum are part of the parahippocampal gyrus (old name, hippocampal gyrus). Color the cut surface of Ammon's horn, the dentate gyrus, and the subiculum.

The Internal Organization of the Hippocampal Formation Figure 52

This figure shows some internal connections within the hippocampal formation. The major source of afferent fibers to the hippocampus is the entorhinal cortex. Neuron "1" and neuron "2" in the medial and lateral entorhinal cortex, respectively, send axons that extend into the dentate gyrus where they synapse upon the dendrites of granule cells.

The granule cells, which are the most conspicuous feature of the dentate gyrus, give rise to mossy fibers that end as giant boutons upon the proximal portion of the apical dendrites of the pyramidal neurons. In the drawing only granule cells "3" and "4" are shown with mossy fiber axons that synapse upon the dendrites of pyramidal cells "5", "6", "7" and "8".

Notice that the axons of the pyramidal cells are directed towards the ventricular surface where most of them form the alveus. The fibers in the alveus group together into the fimbria which is the beginning of the fornix. Note that the pyramidal cell "8" gives off a recurrent collateral axon that synapses upon both the proximal axon and apical dendrites of other pyramidal cells. Locate basket cell "9" in the lower right and note that its axon courses in the pyramidal cell layer of Ammon's horn where its fibers end by making "baskets" about the perikarya of several pyramidal cells. Note how the alveus and fimbria are made; they are essentially aggregates of myelinated axons.

Fibers "10", "11" and "12" represent another source of afferent fibers which come by way of the cingulum. Fiber "10" ends in parasubiculum, fiber "11" in the subiculum, and fiber "12" in the hippocampus proper.

Directions: Starting with neuron "1", color and trace the course of neurons "1-12."

Olfactory Pathways Figure 53

Olfaction or the sense of smell begins at the top of the nasal cavities (figure A) with the olfactory receptor which is a bipolar neuroepithelial cell with chemosensitive cilia (1) projecting from its free surface. These cilia are bathed in a serous fluid that covers the olfactory mucosa. The cilia are stimulated by the various odor-causing molecules that become dissolved in the serous fluid.

The axons of the olfactory receptors (5) group together in small bundles and form the olfactory nerve which is cranial nerve I. These bundles of axons penetrate the cribriform plate of the ethmoid bone, enter the olfactory bulb, branch extensively, and form whorl-like synaptic formations with the apical dendrites of mitral cells (6). These whorl-like formations are the olfactory glomeruli.

The axons of the mitral cells (8) carry the stimuli backwards in the olfactory tract until they reach the olfactory trigone. All of the mitral cell axons (second order fibers) turn into the lateral olfactory stria (lat olf st). Some mitral cell axons synapse upon scattered neurons in the olfactory tract which comprise the anterior olfactory nucleus. Axons of the anterior olfactory nucleus (third order fibers) may pass into either the medial olfactory stria (med olf st) (9) or into the lateral olfactory stria. Some fibers of the medial olfactory stria enter the anterior commissure, cross to the opposite side, enter the opposite olfactory tract, proceed forward as efferent fibers, and finally end in the opposite olfactory bulb.

In addition to axons of mitral cells, the olfactory tract contains the axons of tufted neurons (not shown) and efferent fibers (not shown).

Fibers within the lateral olfactory stria encircle the limen (Latin, threshold) of the insula and travel from the base of the frontal lobe to the temporal lobe (seen best in the lower figure E). They end in either the gyrus semilunaris (gyr simil) (10), gyrus ambiens (11), or the cortical nucleus of the amygdaloid complex (12). The gyrus semilunaris is the equivalent of the periamygdaloid cortex in animals, and the gyrus ambiens corresponds to the prepiriform cortex in animals. These two gyri, the semilunaris and ambiens, are quite small in the human and comprise the primary olfactory cortex (olf cortex, upper right figure).

Recent work has shown that the entorhinal area, 28 of Brodmann, long regarded as the secondary olfactory cortex and not the recipient of direct fibers from the olfactory bulb, does in fact receive primary olfactory fibers.

Humans are not "big smellers". They rely more on vision than on olfaction. Therefore, the human olfactory bulb, tract, stria, and olfactory cortex are quite small. Humans are microsmatic (Greek, micros, small, osme, odor, smell) or "small smellers". Most mammals are "big smellers" or macrosmatic animals.

The olfactory stimuli proceed in the following sequence (figure A on left): chemosensitive cilia (1), olfactory vesicle (2), dendrite of olfactory receptor (3), perikaryon of olfactory receptor (4), axon of olfactory receptor (5), penetration of both cribriform plate and overlying olfactory bulb, synapse with mitral cell dendrites in olfactory glomerulus (6), mitral cell perikaryon (7), mitral cell axon in olfactory tract (8), olfactory trigone, medial and lateral olfactory striae (central two figures C, D), and termination in either the gyrus semilunaris (10), the gyrus ambiens (11), or the cortical nucleus of the amygdaloid complex (also lower right figure E).

Directions: Starting with the chemosensitive cilia (1) in the lower left, color the olfactory pathways yellow.

The Thalamus 1 Anterior and Medial Nuclei   Figure 54

The thalamus (Greek, bedroom*) forms the major portion of the diencephalon. The remainder of the diencephalon consists of the hypothalamus, subthalamus, and epithalamus.

A sheet of white matter, the internal medullary lamina divides the thalamus into three nuclear masses: the anterior (Ant), lateral (L), and medial (M).

In figure A color each of the nuclear masses anterior (Ant), lateral (L), and medial (M) a different color, but leave the internal medullary uncolored.

The lateral and anterior surface of the thalamus is covered by a thin sheet of nuclear material, the thalamic reticular nucleus (figure B). The thalamic reticular nucleus is separated from the rest of the thalamus by another sheet of white matter, the external medullary lamina. Color the thalamic reticular nucleus in figure B.

Figure C shows the internal medullary lamina. Note that it splits anteriorly (A) to partially enclose the anterior thalamic nuclei (Ant). It also splits within the thalamus to enclose certain other nuclei. These nuclei that are entirely enclosed within the internal medullary lamina are called the intralaminar nuclei of which the centromedianum nucleus (CM) is the largest. Color the internal medullary lamina in figure C.

Figure D. The anterior nuclear group (Ant) contains three nuclei of which the anteroventral nucleus is the largest. Color the anterior nucleus group (Ant) the same color as used in figure A.

Figure E shows the medial nuclear masses of the thalamus. These include the dorsomedial nucleus (DM) and the midline nuclei (Mid) which lie most medial and line the walls of the third ventricle and form the interthalamic adhesion (ITA) when present. The habenula (H) is actually part of the epithalamus and not considered a thalamic nucleus. The dashed lines in figures E and F indicate planes of section into which figures G and H have been cut.

Color the medial group nuclei (DM and Midline).

---

*The reason the large diencephalic nuclear mass is called "thalamus" which is the Greek word for "bedroom" or "inner chamber" is due to faulty translation as well as to erroneous anatomy. The ancients believed the nerves to be hollow and to carry particles of sensation to hollow chambers or "thalami" deep within the brain. Presumably these chambers were the ventricles in the brain. Later when anatomists discovered that the optic nerve ended in the thalamus and not into the ventricles, they named this the "optic thalamus" apparently not realizing that originally "thalamus" meant "hollow". Still later the "optic" was dropped and the large cellular mass in the diencephalon became known simply as the "thalamus".

The Thalamus 2 Lateral and Posterior Nuclei   Figure 55

Figure A show the thalamic nuclei that lie lateral and posterior to the internal medullary lamina.

The lateral and posterior nuclear mass has been divided into a ventral tier (figure B) and a dorsal tier (figure C).

Figure B. The ventral tier nuclei consist of a ventral anterior nucleus* (VA), a ventral lateral nucleus (VL), a ventral intermediate nucleus (VI), and a ventral posterior nucleus that is subdivided into a ventral posterolateral nucleus (VPL) and a ventral posteromedial nucleus (VPM).

Figure C. The dorsal tier nuclei consist of a lateral posterior nucleus (LP), a lateral dorsal nucleus (LD), and the pulvinar (Latin, cushion, pillow) (Pul).

The lateral geniculate body (LGB) and medial geniculate body (MGB) are located beneath the pulvinar (figure D).

The ventral tier nuclei VI, VPL, and VPM are shown separated in figure E. The nuclei VPL and VPM are important somatosensory integrative and relay nuclei and were discussed with the somatosensory pathways (figures 12 and 13).

Figure F shows the right thalamus from the posterior, lateral, and superior aspect. Note the position of the medial geniculate body (MGB) and lateral geniculate body (LGB) beneath the pulvinar.

Starting with figure E in the lower left, color the nuclei VPL and VPM in each figure. Also color the medial and lateral geniculate bodies in figures D and F.

Figure E also includes the centromedianum nucleus (CM), an intralaminar nucleus, and indicates its relationship to nuclei VPL and VPM.

---

*Not to be confused with the anteroventral nucleus in the anterior nuclear group.

The Cerebral Cortex Figure 56

The bumps or convolutions of the cerebral cortex are called <u>gyri</u> (singular, <u>gyrus</u>; from the Greek <u>gyros</u>, circle). The gyri are separated from each other by grooves or furrows called <u>sulci</u> (singular, <u>sulcus</u>; Latin for furrow). Large clefts such as the longitudinal fissure between the two cerebral hemispheres are called <u>fissures</u> (Latin, <u>fissura</u>, cleft). The drawing shows a few of the cortical areas in which certain functions have been localized. The numbers are those of Brodmann who divided the cerebral cortex up into some 50 areas on the basis of different cell types. The central sulcus (also called the fissure of Rolando) separates the frontal lobe of the cerebral hemisphere from the parietal lobe. The lateral fissure (also called fissure of Sylvius) separates the temporal lobe from the frontal and parietal lobes. Note that this separation is only partial and that the temporal lobe is joined medially to the frontal lobe and posteriorly to the parietal and occipital lobes. If the lateral (sylvian) fissure is pulled apart additional hidden cortex is revealed. This hidden cortex is the <u>insula</u> (Latin, island). The cortex that covers the insula belongs to the frontal, parietal and temporal lobes. The central sulcus lies between the anterior primary motor cortex (area 4 of Brodmann) and the posterior primary somatosensory cortical areas (3, 1, 2 of Brodmann). Note that 3, 1, and 2 are three parallel strips on the postcentral gyrus. Note also that Brodmann's area 4 is wide superiorly and becomes a narrow strip inferiorly as it approaches the lateral or Sylvian fissure. Note in the lower figure which is a medial view that areas 4 and 3, 1, 2 overlap onto the medial surface of the cortex.

<u>Directions</u>: Color each of the cortical areas discussed using a different color or shade for adjacent areas.

Area 44 on the posterior and inferior part of the left frontal lobe is Broca's area and is responsible for the muscular control in normal speech. Damage to area 44 results in <u>expressive aphasia</u> in which the victim is able to comprehend what he hears, but lacks the motor control of his vocal apparatus and cannot express himself intelligibly. Area 41 on the superior surface of the temporal lobe is the primary auditory cortex which receives the auditory projection from the medial geniculate body. Areas 42 and 22 are the auditory association areas. Areas 41 and 42 on the left hemisphere together comprise Wernicke's area. Damage to Wernicke's area results in <u>receptive aphasia</u> or <u>Wernicke's aphasia</u> in which the victim cannot understand what is being spoken to him. He himself can speak but his own speech tends to be meaningless even though he is able to form and express words. There seems to be a pathological tendency to use the "wrong word". Areas 39 and 40 which lie above the sylvian fissure on parietal lobe are essential for the ability to read and comprehend written language. These areas, namely 44, 41, 42, 39, and 40, which are grouped around the posterior part of the sylvian fissure on the left hemisphere (collectively they are referred to as the <u>perisylvian region</u>) are essential for the normal expression, reading and understanding of language.

The <u>occipital cortex</u> at the posterior pole of the cerebral hemisphere was divided by Brodmann into three areas, 17, 18, and 19. Area 17 is the primary visual cortex and receives the optic radiation from the lateral geniculate body. Note in the lower figure that most of area 17 is on the medial surface of the cerebral cortex. Area 17 is also called the <u>striate cortex</u> because of a prominent white line in its fourth layer. This line is the line or stria of Gennari and its presence readily identifies area 17. Area 17 is responsible for localizing objects in space and for the "solid" or three-dimensional appearance of the world about us.

One half of each visual and retinal field will end in each area 17. The right visual field and left retinal field of both eyes will end in the left visual cortex. The left visual field and right retinal field of both eyes will end in the right visual cortex. The two images from each eye are somehow fused by the occipital cortex into a "solid" or a stereoscopic image. Color is perceived in area 17. Stimulation of area 17 either electrically by a wire applied directly to the cortex or by a focal epileptic seizure has brought forth simple visual hallucinations in which the individual has reported "seeing" flashes of lights, bright points, stars, color flashes, and geometric forms such as circles, squares, and hexagons. Electrical stimulation of the surrounding areas 18 and 19 has evoked more complex visual hallucinations in which the patient "sees" animals, people and objects. These hallucinations may be so "real" that the patient actually believes he is seeing them. He may react to them with fear or he may think they are amusing and laugh at them. Hallucinations may be defined as false perceptions that are somehow induced by an abnormal neuronal discharge in the central nervous system. Areas 18 and 19 are the visual association areas and are responsible for the interpretation and understanding of what one sees. They integrate the present visual perception with one's past experience and thus impart a significance and interpretation to the visible world about us. Damage to areas 18 and 19 will result in visual agnosia in which the victim is able to "see" objects in the sense that visual impulses reach the conscious level, but he does not understand what he sees. A gold watch held in front of the patient is merely "something shiny." But if the watch is held close enough for its ticking to be heard, he will recognize it instantly as a watch. The patient may not even be able to recognize his own relatives visually. They are just "people" or "someone" until they speak and they are recognized by their voices. Damage to the areas in front of area 19 in both the parietal and temporal cortex may give rise to distortions of perception called illusions. Illusions are alterations of perceptions and differ from hallucinations in that hallucinations are completely false perceptions with no relation between the mental event that is being "experienced" and the external world; whereas illusions are the mental or emotional "stretching, pulling, coloring" or distortion of sensory stimuli that are based upon actual events and objects in the world about us. Abnormal functioning of the cortical areas immediately in front of the visual cortex may cause visual illusions in which objects may appear larger than they actually are (macropsia), or smaller (micropsia), or appear to move away from or towards the patient. There may be a loss of stereoscopic vision in which everything looks flat and two-dimensional. Or the visual image may persist (palinopsia).

The parietal lobe extends from the occipital lobe posteriorly to the central sulcus anteriorly. Its most anterior gyrus is the postcentral gyrus that contains Brodmann's areas 3, 1, and 2. These three cortical areas constitute the primary somatosensory cortex upon which the opposite side of the body is projected, by way of nucleus VPL and nucleus VPM, in an orderly topographical pattern. Areas of the body with the greatest sensitivity such as the finger tips and lips have the largest cortical areas of representation. On the other hand, regions of the body such as the back which is considerably less sensitive have relatively small cortical areas. This sensory projection in the somatosensory cortex is said to comprise the sensory homunculus (Latin, tiny man), and the motor cortex contains the motor homunculus. Within the sensory projection, the face is the nearest the sylvian fissure and the rest of the body is projected in an upside-down manner so that the knee area lies at the highest point where the convex lateral surface turns onto the vertical medial surface upon which the lower leg and foot are projected. The somatosensory cortex is not strictly sensory in that many fibers in the pyramidal tract have their cells of origin in areas 3, 1, 2 and other parietal cortical areas also send large contingents of fibers to the pyramidal tract. Therefore, the somatosensory cortex has been given the designation Sm, large "S" for sensory and small "m" for motor indicating it is predominantly sensory with some motor function. Areas 3, 1, 2 are not the only somatosensory areas; there are believed to be two other such areas. Therefore, areas 3, 1, 2, the primary somatosensory area, are Sm I, and the other two somatosensory areas are Sm II and Sm III.

Electrical stimulation of points within the primary somatosensory cortex by neurosurgeons has evoked feeling such as itching, tingling, numbness, and a desire to move. These sensations are experienced at the body surface and tend to be sharply localized. Similar feelings are brought on by focal epileptic seizures which have their irritative source or focus in or near the primary somatosensory area. The parietal lobe as a whole maintains a mental perception of the body, both in terms of where the body is in relation to its surroundings and where each part of the body is in relation to the others. This mental perception of our bodies is proprioception (Latin, proprius one's own, and captus, comprehension) which is a complex sense that is built up from many simpler sensations such as those from joints, tendon organs, touch receptors, and pressure receptors.

Following World War I Holmes and Head in England examined soldiers who suffered damage to their parietal lobes. Holmes and Head found that whereas their patients were able to sense gross movements of their limbs, they could not tell what direction their foot was being moved by the examiner unless they observed it visually. They were unable to detect small differences in weight. Two points touching their skin were felt as being one point unless the two points were considerably further apart than the distance in normal individuals. They also found that parietal lobe damage severely affected the sense of touch, particularly the ability to use the fingers to identify objects by their shape and texture. This ability is called stereognosis (Greek, stereos, solid and gnosis, knowledge) and, like proprioception, depends upon the synthesis of several sensory modalities such as joint sense, tactile sense, and pressure sense. The inability to recognize objects by their shape is called astereognosis (see figure 14).

The temporal cortex, as already mentioned, contains the primary auditory cortical area 41 and auditory association areas 42 and 22. Focal epileptic seizures that have their focus or point of origin in the primary auditory cortex bring forth simple auditory hallucinations such as the "sound" of "birds flapping their wings" or the "sound" of a "waterfall" or "wind blowing through the trees." Stimulation of the auditory association areas 42 and 22 evokes more complicated auditory hallucinations such as a portion of a conversation that actually occurred in the past. The Canadian neurosurgeon Penfield described one of his epileptic patients who, during stimulation of his auditory association area, "heard" a portion of a song from his past that was so believable that the patient felt he was actually reliving the experience. Observations such as this in which memories seem to be recalled by stimulation of the temporal cortex have linked the temporal lobe strongly to memory storage and recall. Penfield and others have suggested that the temporal lobe retains a permanent neuronal memory of our past experiences, and we use the temporal lobe as an emotional guide for judging our present situation which is continually being conveyed to our consciousness by sensory stimuli. Where these memory traces reside in the temporal lobe is not at all understood. It is not known if the memory traces are stored in the temporal cortex or in deeper temporal structures such as the hippocampus or even within the cores of white matter that run from the interior of the temporal lobe to the interiors of the other lobes of the cerebrum. Rats who have had bilateral lesions of their hippocampal formation take much longer to run through a maze and apparently are unable to learn and remember the pathway even after they have run the course several times. This suggests a definite memory function for motor acts for the hippocampal formation in rats. However, the results from human studies of the hippocampal role in memory are inconclusive. In what has now become a classic study, Kluver and Bucy removed both temporal lobes from animals. They recorded a number of striking changes which comprise the so-called Kluver-Bucy syndrome. After bilateral ablation (removal) of their temporal lobes, monkeys displayed visual agnosia. They were not able to recognize objects by sight but had instead to feel objects with their fingers or put objects in their mouth as babies do. This fingering and insertion of objects into the mouth included nuts and bolts and even live snakes which monkeys normally avoid. There also were marked changes in their eating habits in which fruit-eating Rhesus monkeys ate meat. Bilateral temporal lobectomy left the usually aggressive Rhesus monkeys placid and emotionally flat. They also displayed abnormal sexual behavior in which male monkeys attempted to copulate with chickens.

Temporal lobe epilepsy gives rise to a variety of "sensory-type" seizures most of which have no motor component. These have been described as altered feelings, distorted perception (or illusions), and psychical hallucinations. When they are accompanied by a motor act (often characterized by a smacking of the lips or movements of the facial muscles) they constitute psychomotor seizures. Altered feeling or emotional seizures are overwhelming feelings such as fear or dread. These seizures are probably induced by a seizure in or near the uncus (uncal seizures) which involved the underlying amygdala. Another type of emotional seizure is feelings of euphoria (feeling good or pleasure). There may be feelings of familiarity that convince the individual that he has "been in this same place" or "done this same thing" in the past, even though the individual knows he is not repeating a familiar act. This feeling of familiarity is deja vu and is something that we have all experienced in mild forms ourselves. There may be feelings of remoteness. All of these emotional seizures are believed to involve an abnormal discharge in the temporal lobe.

Distorted perception may be visual (as described with the visual cortex), tactile (objects feel strange), or auditory. Auditory illusion would include sound seeming unusually loud or unreal or a certain word being extremely unpleasant. Penfield has reported a few cases of what he called psychical hallucinations induced by seizures in the temporal cortex.
One of Penfield's patients was a woman who had generalized convulsions that were always preceded by the recall of a very frightening event of her past. She felt herself to be "reliving" this terrifying memory just before the onset of a generalized seizure. As a young girl this woman was frightened when playing in a field with her brothers by a man who approached her and threatened to put her in a sack he was carrying. Nothing more actually happened, but some years later she developed grand mal (generalized) epilepsy and each seizure was always preceded by her "experiencing" this horrible event in which she feels herself to be the little girl being frightened by the man with the sack. Penfield surgically operated upon this woman and found the irritative source of her seizures and hallucinations to be scar tissue on her temporal cortex. He removed this surgically and eradicated the seizures and the physical hallucinations.

The frontal lobe reaches its greatest elaboration in the human where it is the largest of four lobes. Its most posterior part is area 4, the primary motor cortex, that lies just in front of the central sulcus. The primary motor cortex contains a motor representation similar to the sensory representation in areas 3, 1, 2. Parts of the body capable of the most complex movements such as the fingers, lips, and tongue have the largest cortical areas. Electrical stimulation of points within the primary motor cortex causes twitching of small groups of muscles in the opposite side of the body. A Jacksonian fit or seizure is essentially the same thing, that is, a twitching of a small group of muscles caused, in this case, by an irritative source in or near the primary motor cortex. In some cases, the abnormal discharge within the motor cortex does not remain localized but instead spreads throughout the whole primary motor cortex. This is a Jacksonian march and is characterized by the seizure beginning with convulsions in a small area of the body, for example, the thumb. The convulsions then spread to adjoining regions such as the hand, forearm, elbow, and so forth until one-half of the body is undergoing convulsions. Should the convulsions reach the other side of the body, the individual will lose consciousness. The primary motor cortex contains the motor homunculus of the opposite side of the body. The primary motor cortex is not exclusively motor, but contains a considerable somatosensory projection, thus it has been designated Ms I. Large "M" for motor, its major component, small "s" for sensory, its minor component, and "I" because it is the primary motor cortex, there being a supplementary motor cortex on the medial surface of the frontal lobe. Movements can also be elicited by stimulating premotor area 6, but these movements are likely to involve more muscles and require stronger stimuli. Area 8 in front of 6 is another premotor area and contains frontal eye fields. Stimulation here causes horizontal or oblique conjugate movements of the eyes to the contralateral side.

The cortical area in front of the motor and premotor areas, the prefrontal cortex has long been regarded as a "silent area". It was "silent" in the sense that seizures seldom if ever arose from this region; nor would electrical stimulaton reveal anything; neither any movement, as was the case with motor cortex stimulation, nor sensory hallucinations (numbness, tingling, etc.), as with somatosensory stimulation. An intact prefrontal cortex is essential for such higher human functions as the ability to plan ahead and foresee the consequences of the course of action one is contemplating, particularly in terms of people and human response. Whether these human characteristics such as social tact, concern for others, worry, and anxiety actually "reside" in the prefrontal lobe or depend upon the interaction of this part of the brain with other parts for their normal function and expression can only be speculated. What is apparent is that prefrontal damage may profoundly alter one's "personality" so that he becomes socially indifferent (or tactless) and is unable to pursue and complete a complicated or extended course of action. There seems to be an emotional flattening with an imbalance or disruption of intellect with emotions. In the 1940's and 1950's, various forms of psychosurgery were performed upon the prefrontal lobes of patients in mental hospitals. These were the prefrontal lobotomies and leukotomies in which various incisions were made into the underlying white matter of the prefrontal cortex. These operations had at best a very limited success and these were mainly with obsessional psychoses and extreme depression. Advocates of leukotomies claimed effective results when the white matter that connected the dorsal medial thalamic nucleus with the orbital cortex was severed.

At a time when nothing except electroconvulsive therapy seemed to do any good, and even that had limited success, the prefrontal lobotomies appeared to promise new hope. It was claimed that many of the lobotomy patients could return home and lead more normal lives and, indeed, many lobotomy patients were sent home. The practice of lobotomies became standard procedure in mental hospitals in this country and Europe and tens of thousands of patients were lobotomized. Newer and simpler forms of the operation and instruments were devised including a "cosmetic approach" that left practically no scar or outward signs of surgery and did not require the head to be shaved. This was a great relief to the women patients who were often more concerned about the heads being shaved than the operation itself. The "cosmetic approach" required the leukotome to be inserted behind the eye lid and then to be forced up through the orbital plate of the frontal bone into the underside of the brain to a depth, as one report recommends, of some 7 centimeters. Allowing for eye lid, bone and subarachnoid space, this meant some 5 centimeters (about 2 inches) of metal was shoved into the orbital cortex and deeper white matter. The leukotome could then be moved in an arc from side to side somewhat like a scythe cutting wheat and a great swath would be cut through the fibers that connect the prefrontal cortex with the more posterior parts of the brain.

With the discovery of tranquillizing and depressant drugs, prefrontal lobotomies were no longer necessary and the practice was discontinued, although even today a limited type of leukotomy is sometimes found to be the only measure that gives relief from intractable pain in terminally ill patients. A "cingulotomy," in which the association fibers beneath the anterior cingulate gyrus are cut, seems to abolish or greatly reduce the overwhelming intensity of pain. Surprisingly, after such an operation, the patient is still conscious of the pain, but it no longer seems to bother him.

It is now realized that the comprehension and production of language is located in the great majority of cases in the left cerebral hemisphere. The left cerebral hemisphere is regarded as the dominant hemisphere and the right hemisphere as the non-dominant hemisphere. More than 95% of right-handed individuals have dominant left hemispheres. Even the majority of the left-handed people have dominant left hemispheres, although left-handed people are more likely than right-handed to have the right dominant hemispheres or have some language function in both hemispheres. There are also actual anatomical differences between the two hemispheres, especially around the sylvian fissure which is longer on the left side with considerably more cortical tissue around the left fissure than on the right.

Not only is the left hemisphere dominant for language, but also for mathematical ability and the ability to solve problems in a logical sequential manner. The right hemisphere appears to be superior in tasks involving spatial relations, drawing, recognizing human faces, and in musical skills. The right hemisphere also solves problems, but rather than doing this in a logical step-by-step manner, the right hemisphere solves problems in a comprehensive holistic manner. The corpus callosum which is an enormous commissure of about 300 million fibers connects the two cerebral hemispheres and informs one hemisphere what is happening in the other hemisphere. Most callosal fibers connect homologous (or mirror image) sites, but a considerable number end in non-homologous (non-mirror image) sites.

The Strange Case of Phineas P. Gage   Figure 57

Phineas P. Gage was a railroad worker who in 1848 survived an accidental explosion in which a 3 1/2 foot steel bar was blown through his head. A considerable part of his brain was destroyed including the left frontal lobe and possibly the right frontal lobe as well.

The profound personality and behavioral changes that Gage displayed following the accident were recorded by Dr. John M. Harlow who treated Gage after his injury. Prior to his accident, Gage was a normal, well-liked, young man who had never been sick a day in his life. Following this mishap, Harlow wrote that Gage was stubborn, lacking in sound judgment, inconsistent in his plans, and inconsiderate towards others.

Because these traits appeared after a good part of Gage's frontal lobes were lost, scientists reasoned that these human qualities, namely reasonableness, ability to plan ahead and consistently pursue a long-range course of action, and consideration for others "resided in" the frontal lobes or, at least depended upon intact frontal lobes for their normal expression.

A century later, certain "unfortunate human qualities" such as extreme anxiety and obsessive-compulsive behavior were also believed to reside in the frontal areas, more precisely the "prefrontal" areas, and this served as the rationale for the "psychosurgical" prefrontal lobotomies mentioned earlier.

Gage lived for twelve and a half more years and when he died in San Francisco in 1861, Harlow was not able to obtain the brain and, therefore, was not able to assess the actual amount of brain damage. However, some years later, Harlow did obtain Gage's skull and from his published illustrations, it appears that the metal bar passed through the left frontal lobe and that the right frontal lobe may have been spared.

It is perhaps germane in light of this to mention a few additional behavioral quirks noted by Gage's mother. She told Harlow that after the accident Phineas would "entertain his little nephews and nieces with the most fabulous recitals of his wonderful feats and hair-breadth escapes without any foundation except in his fancy." He conceived a great fondness for pets and souvenirs, especially for children, horses and dogs - only exceeded by his attachment for his tamping iron which was his constant companion during the remainder of his life.

It is tempting to ask (and this is merely speculation): Could these affectionate child-like traits described by his mother possibly be the result of Gage's right frontal lobe acting on its own, without any dominance or restraint from the left frontal lobe?

A brief summary of Harlow's paper* on the case follows:

Phineas P. Gage was a foreman of a gang of railroad workers constructing a railway track near Cavendish, Vermont. On September 13, 1848, while in the act of tamping in an explosive charge of gun powder, his attention was diverted to his men who were working in a pit behind him. He turned his head and at the same time dropped his tamping iron which "struck fire upon a rock causing the gun powder to explode and shooting the iron bar "completely through his head." The iron bar was later found by his men, "smeared with blood and brain."

"The patient was thrown upon his back by the explosion and gave a few convulsive motions of the extremities, but spoke in a few minutes. His men (with whom he was a great favorite) took him in their arms" and carried him to his hotel where Gage was attended by two physicians, including Dr. John Harlow who later wrote about the case.

Dr. Harlow found Gage to be "perfectly conscious" and "bearing his suffering with firmness." He directed Dr. Harlow's attention to the hole in his cheek saying "The iron entered there and passed through my head." The iron had apparently entered the left side of the face immediately in front of the angle of the mandible, passed through the back of the left orbit (Gage lost sight in his left eye), then through the left frontal lobe of the brain, and emerged from the top of the skull in the midline at the junction of the frontal bone and two parietal bones.

In the days immediately following the accident, Dr. Harlow held no hope for Gage's recovery. Dr. Harlow wrote that "friends and attendants are in hourly expectancy of his death, and have his coffin and clothes in readiness to remove his remains to his native place in Lebanon, New Hampshire...." One of his attendants implored me not to do anything more for him as it would only prolong his sufferings - that if I would only keep away and let him alone, he would die."

However, Dr. Harlow continued to treat Gage, dressing and cleaning his wounds three times a day and bathing the face and head with ice water. "With a pair of curved scissors I cut off the fungi which were sprouting out from the top of the brain and filling the opening, and made free application of caustic to them."

On the thirty-second day, Harlow recorded, "Progressing favorably....Remembers passing and past events correctly, as well before as since the injury. Intellectual manifestations feeble, being exceedingly capricious and childish, but with a will as indomitable as ever; is particularly obstinate, will not yield to restraint when it conflicts with his desires."

A month later, Harlow wrote that Gage continued to improve but "is impatient of restraint, and could not be controlled by friends."

By the following April of 1849, Dr. Harlow is convinced that Gage has recovered his physical health. "He has no pain in head, but says it has a queer feeling which he is not able to describe. ...His contractors, who regarding him as the most efficient and capable foreman in their employ previous to his injury, considered the change in his mind so marked that they could not give him his place again. The equilibrium or balance, so to speak, between his intellectual faculties and animal propensities, seems to have been destroyed. He is fitful, irreverent, indulging at times in the grossest profanity (which was not previously his custom), manifesting but little deference for his fellows, impatient of restraint or advice when it conflicts with his desires, at times pertinaciously obstinate, yet capricious and vacillating, devising many plans for future operation, which are no sooner arranged than they are abandoned in turn for others appearing more feasible. A child in his intellectual capacity and manifestations, he has the animal passions of a strong man. Previous to his injury though, untrained in the schools, he possessed a well-balanced mind....In this regard, his mind was radically changed, so decidedly that his friends and acquaintances said he was 'no longer Gage.'"

In spite of the profound behavioral and personality changes, Gage lived an independent and productive life in his remaining twelve and a half years. He worked with horses in New Hampshire and then spent eight years in Chile running a stage coach line. In 1859, his health began to fail and he left Chile for San Francisco where his mother and sister were then living. His health improved after reaching

San Francisco and he worked there for a farmer. However, his health once again began to fail and he died on May 21, 1861 after a series of epileptic convulsions.

As already mentioned, Dr. Harlow later obtained both the skull of Gage and the tamping iron which he donated to the Medical Museum at Harvard University.

It was a miracle to Dr. Harlow and physicians of that time that anyone could survive and recover from such a horrible accident and, indeed, many physicians refused to believe the story when they first heard it.

In his lecture to the Massachusetts Medical Society on June 3, 1866, Dr. Harlow demonstrated the skull and the tamping iron to his audience and concluded with these words, "I can only say, in conclusion, with good old Ambrose Pare, I dressed him, God healed him."

---

*Harlow, J. M., 1868. Recovery from the Passage of an Iron Bar Through the Head. Publications of the Massachusetts Medical Society. 2. 320-347.

☐ 1 Efferent
☐ 2 Afferent

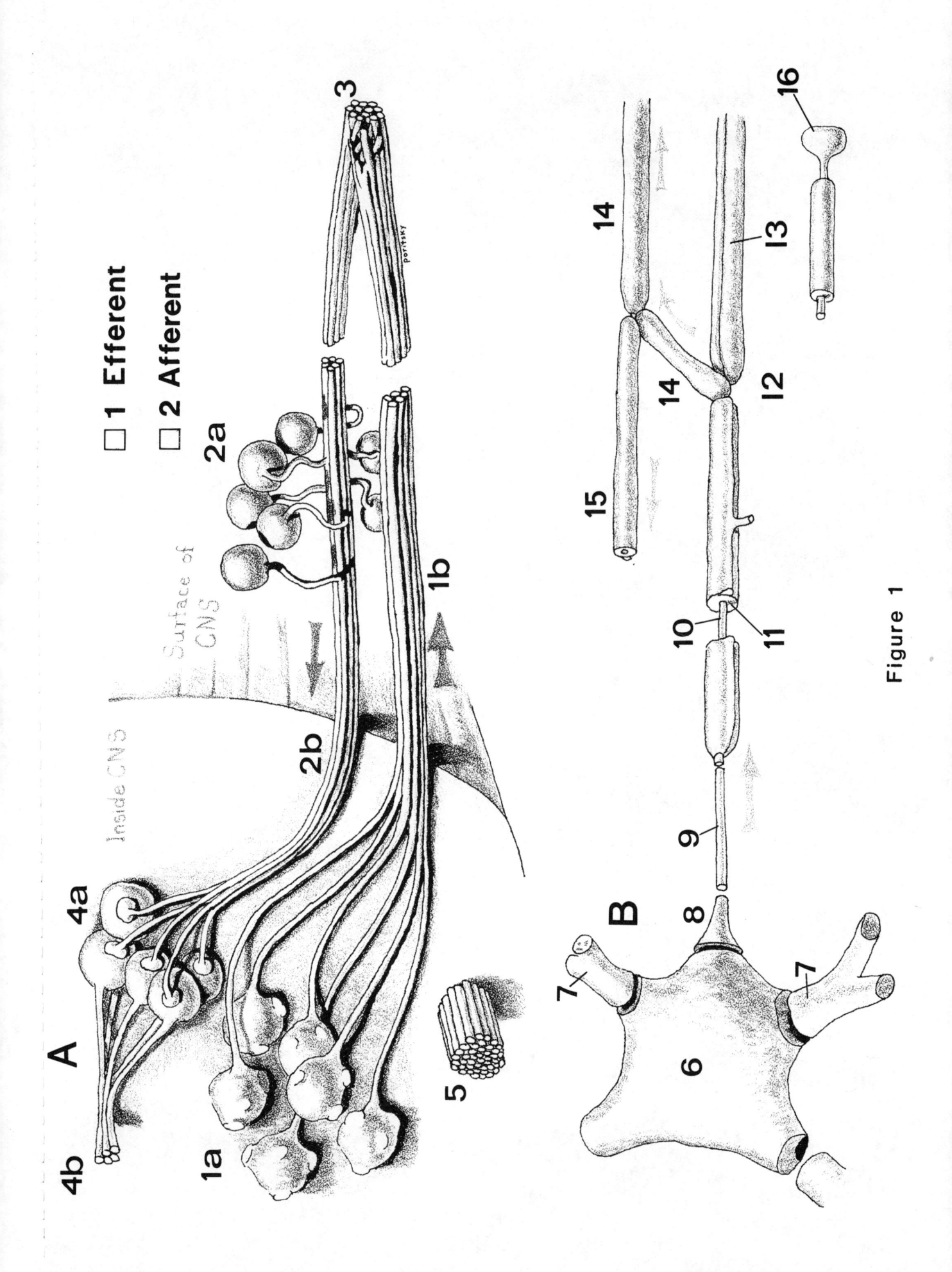

Figure 1

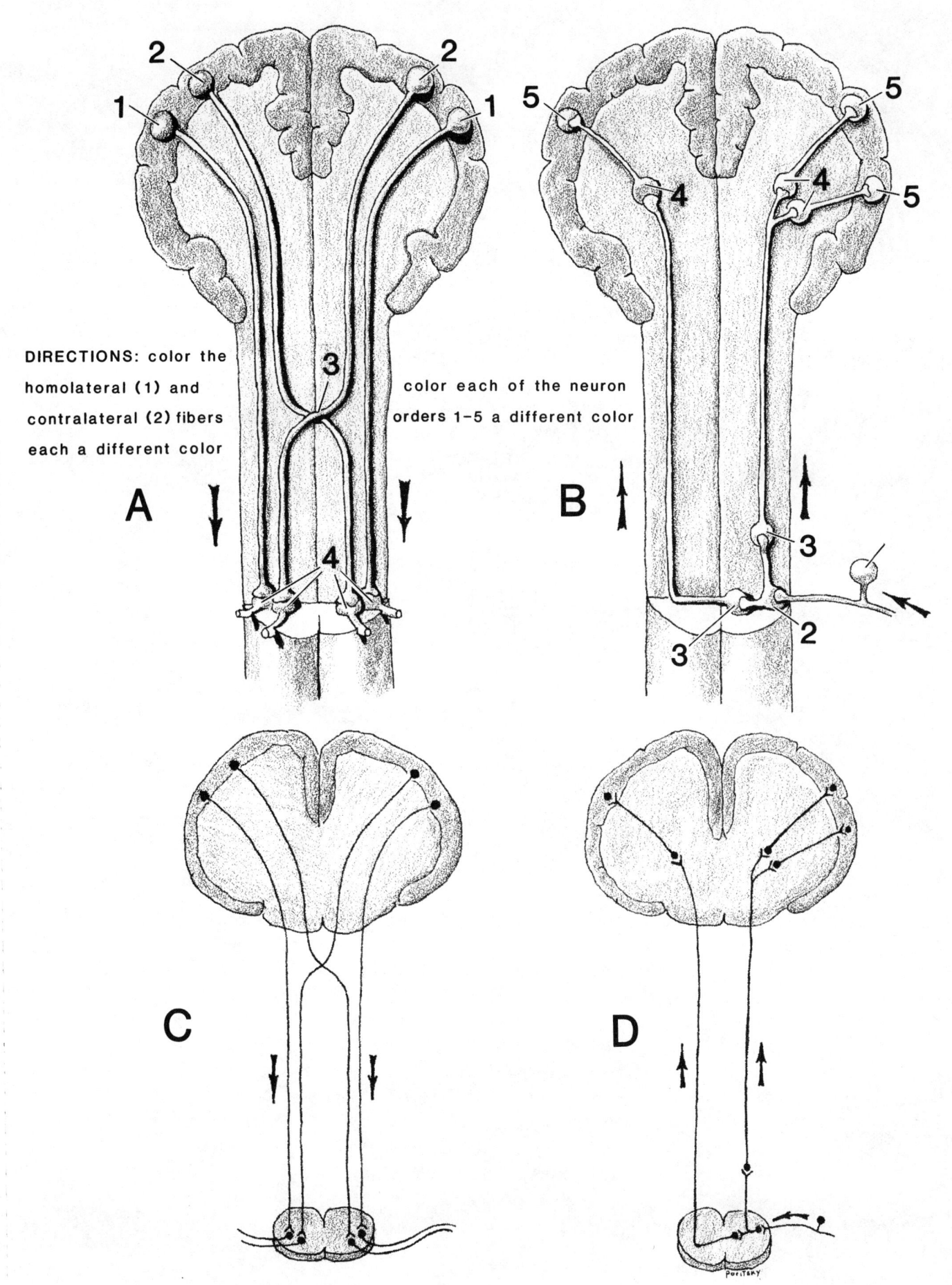

DIRECTIONS: color the homolateral (1) and contralateral (2) fibers each a different color

color each of the neuron orders 1-5 a different color

Figure 2

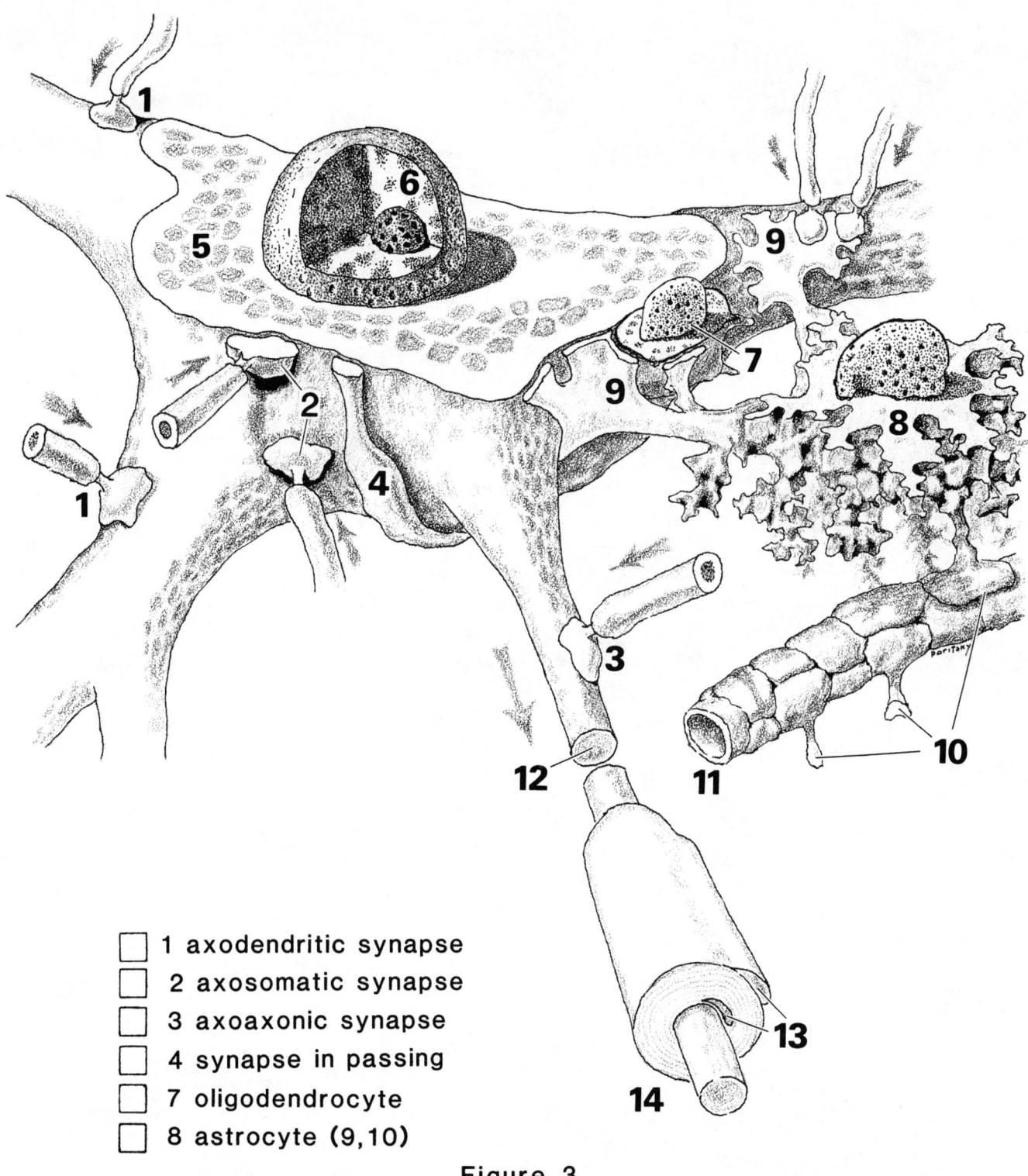

☐ 1 axodendritic synapse
☐ 2 axosomatic synapse
☐ 3 axoaxonic synapse
☐ 4 synapse in passing
☐ 7 oligodendrocyte
☐ 8 astrocyte (9,10)

Figure 3

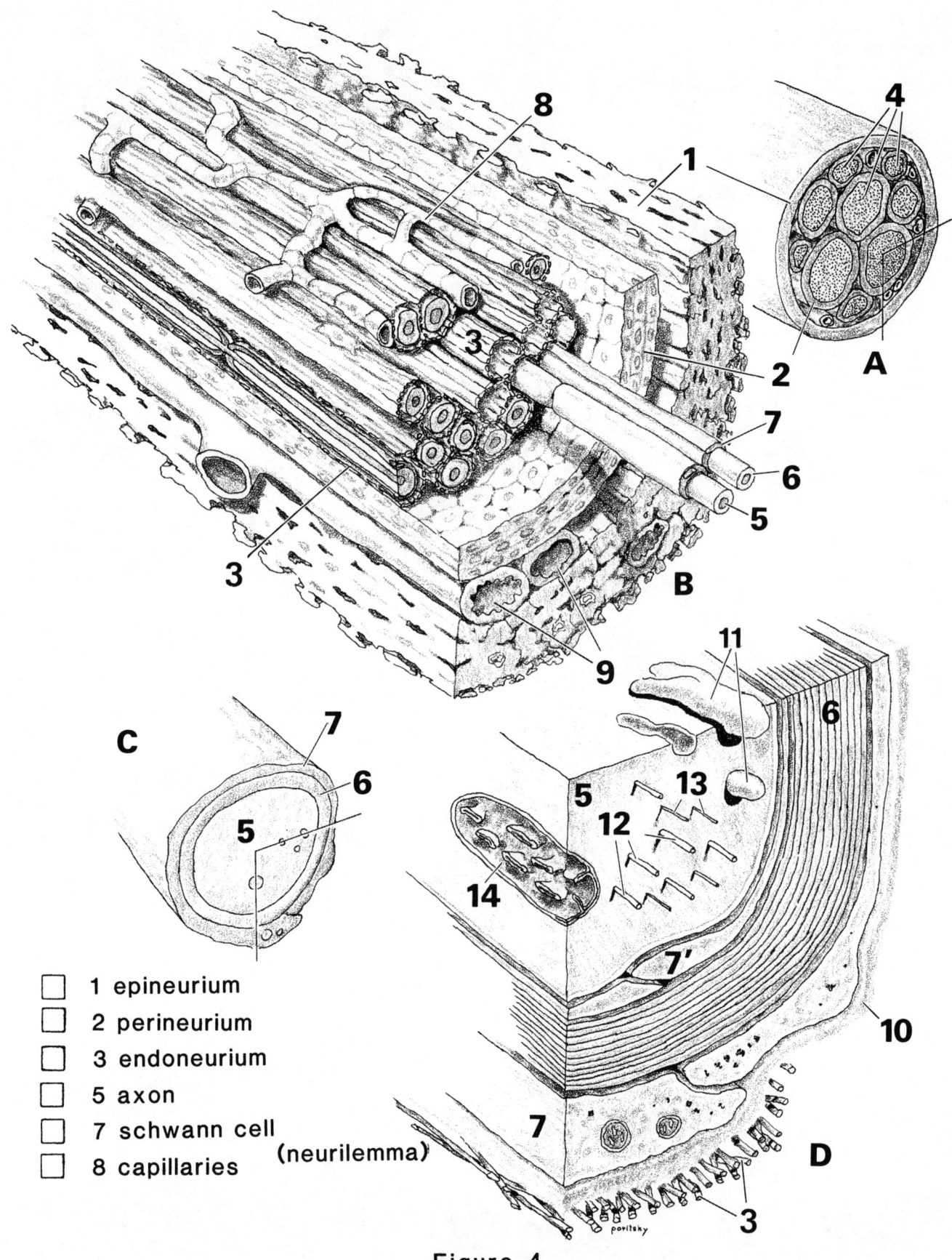

- 1 epineurium
- 2 perineurium
- 3 endoneurium
- 5 axon
- 7 schwann cell (neurilemma)
- 8 capillaries

Figure 4

COLOR EACH OF THE FOLLOWING *

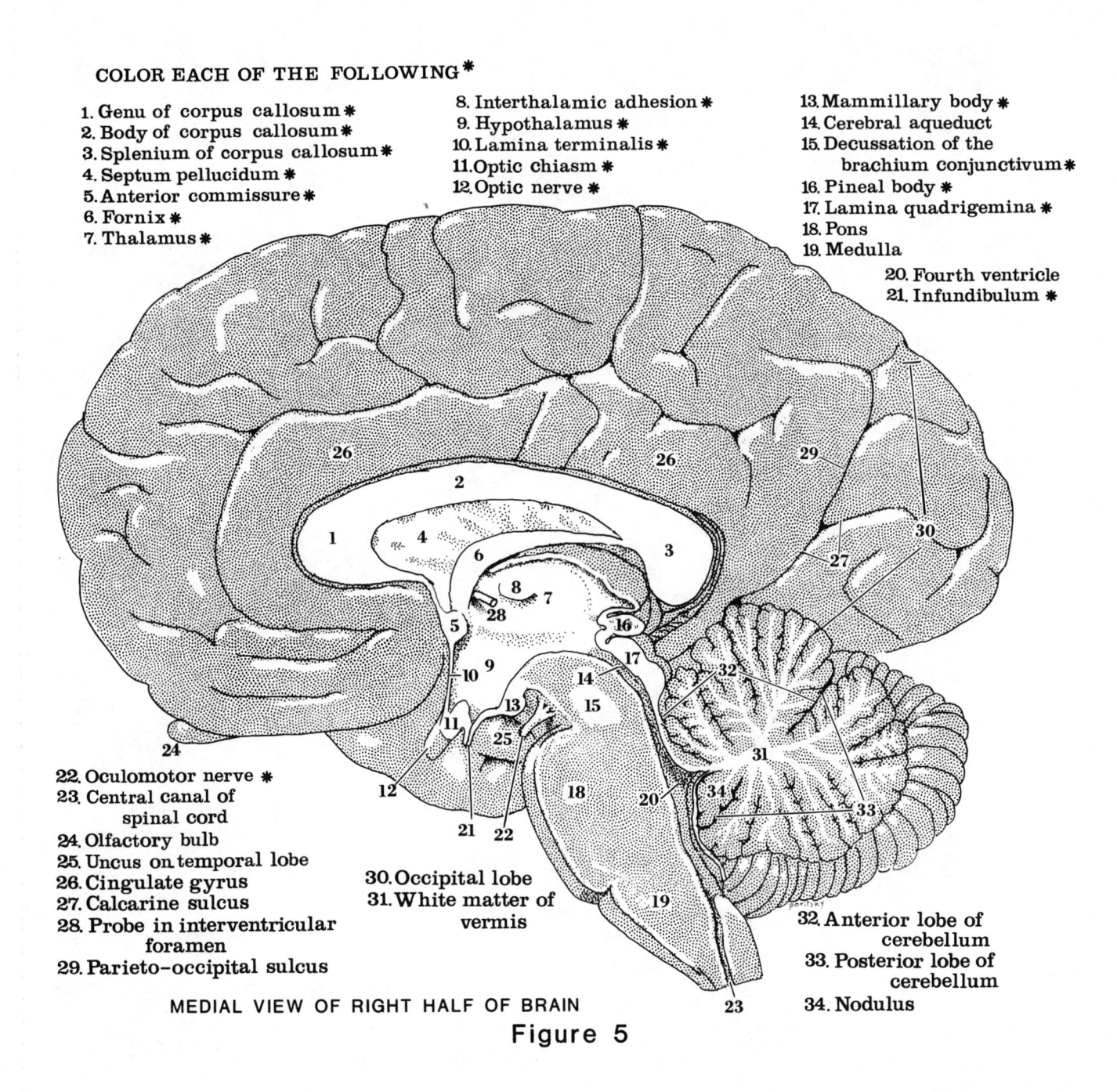

1. Genu of corpus callosum *
2. Body of corpus callosum *
3. Splenium of corpus callosum *
4. Septum pellucidum *
5. Anterior commissure *
6. Fornix *
7. Thalamus *
8. Interthalamic adhesion *
9. Hypothalamus *
10. Lamina terminalis *
11. Optic chiasm *
12. Optic nerve *
13. Mammillary body *
14. Cerebral aqueduct
15. Decussation of the brachium conjunctivum *
16. Pineal body *
17. Lamina quadrigemina *
18. Pons
19. Medulla
20. Fourth ventricle
21. Infundibulum *

22. Oculomotor nerve *
23. Central canal of spinal cord
24. Olfactory bulb
25. Uncus on temporal lobe
26. Cingulate gyrus
27. Calcarine sulcus
28. Probe in interventricular foramen
29. Parieto-occipital sulcus
30. Occipital lobe
31. White matter of vermis
32. Anterior lobe of cerebellum
33. Posterior lobe of cerebellum
34. Nodulus

MEDIAL VIEW OF RIGHT HALF OF BRAIN

Figure 5

INFERIOR VIEW OF BRAIN

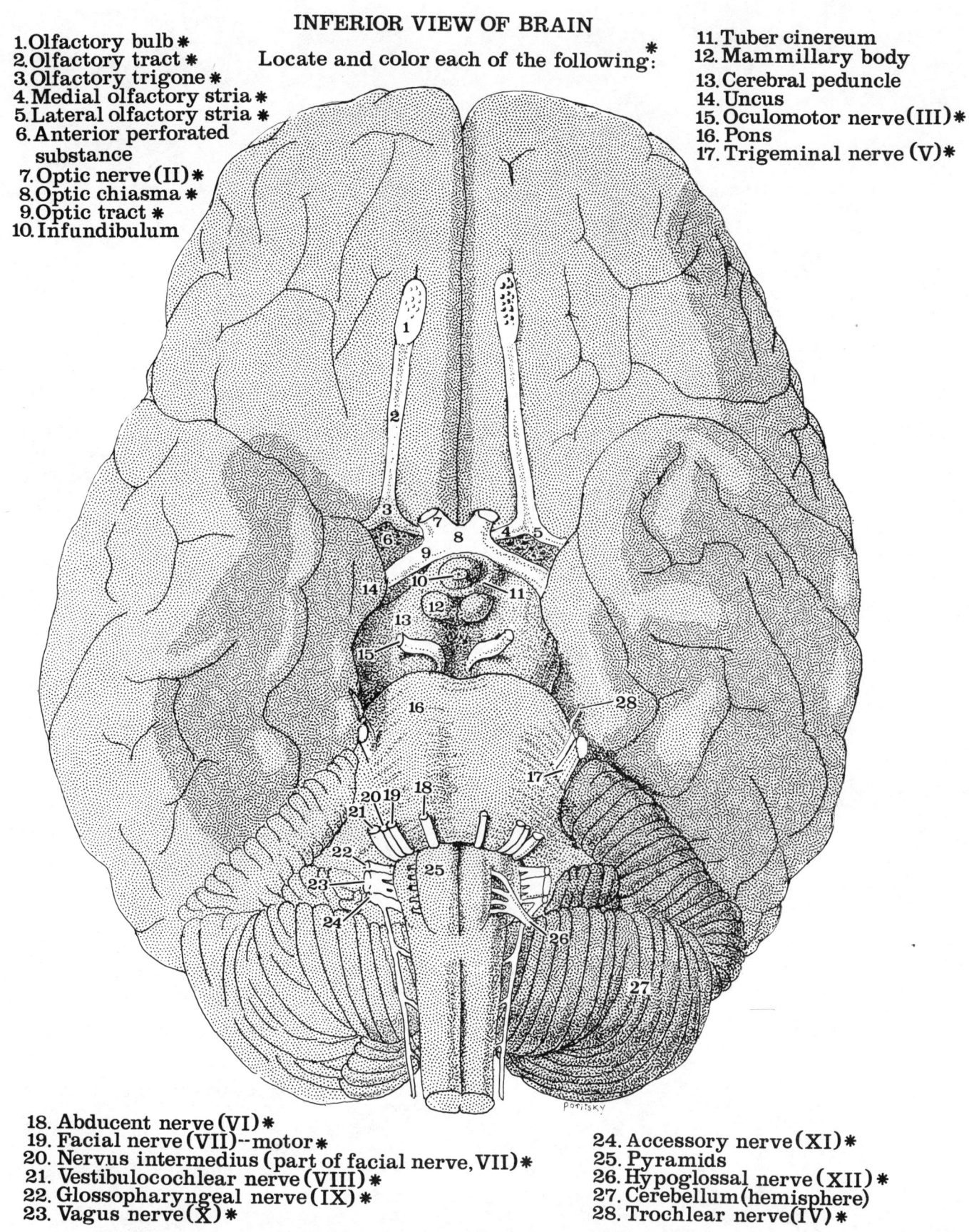

Locate and color each of the following:

1. Olfactory bulb*
2. Olfactory tract*
3. Olfactory trigone*
4. Medial olfactory stria*
5. Lateral olfactory stria*
6. Anterior perforated substance
7. Optic nerve (II)*
8. Optic chiasma*
9. Optic tract*
10. Infundibulum
11. Tuber cinereum
12. Mammillary body
13. Cerebral peduncle
14. Uncus
15. Oculomotor nerve (III)*
16. Pons
17. Trigeminal nerve (V)*
18. Abducent nerve (VI)*
19. Facial nerve (VII)--motor*
20. Nervus intermedius (part of facial nerve, VII)*
21. Vestibulocochlear nerve (VIII)*
22. Glossopharyngeal nerve (IX)*
23. Vagus nerve (X)*
24. Accessory nerve (XI)*
25. Pyramids
26. Hypoglossal nerve (XII)*
27. Cerebellum (hemisphere)
28. Trochlear nerve (IV)*

Figure 6

Figure 7

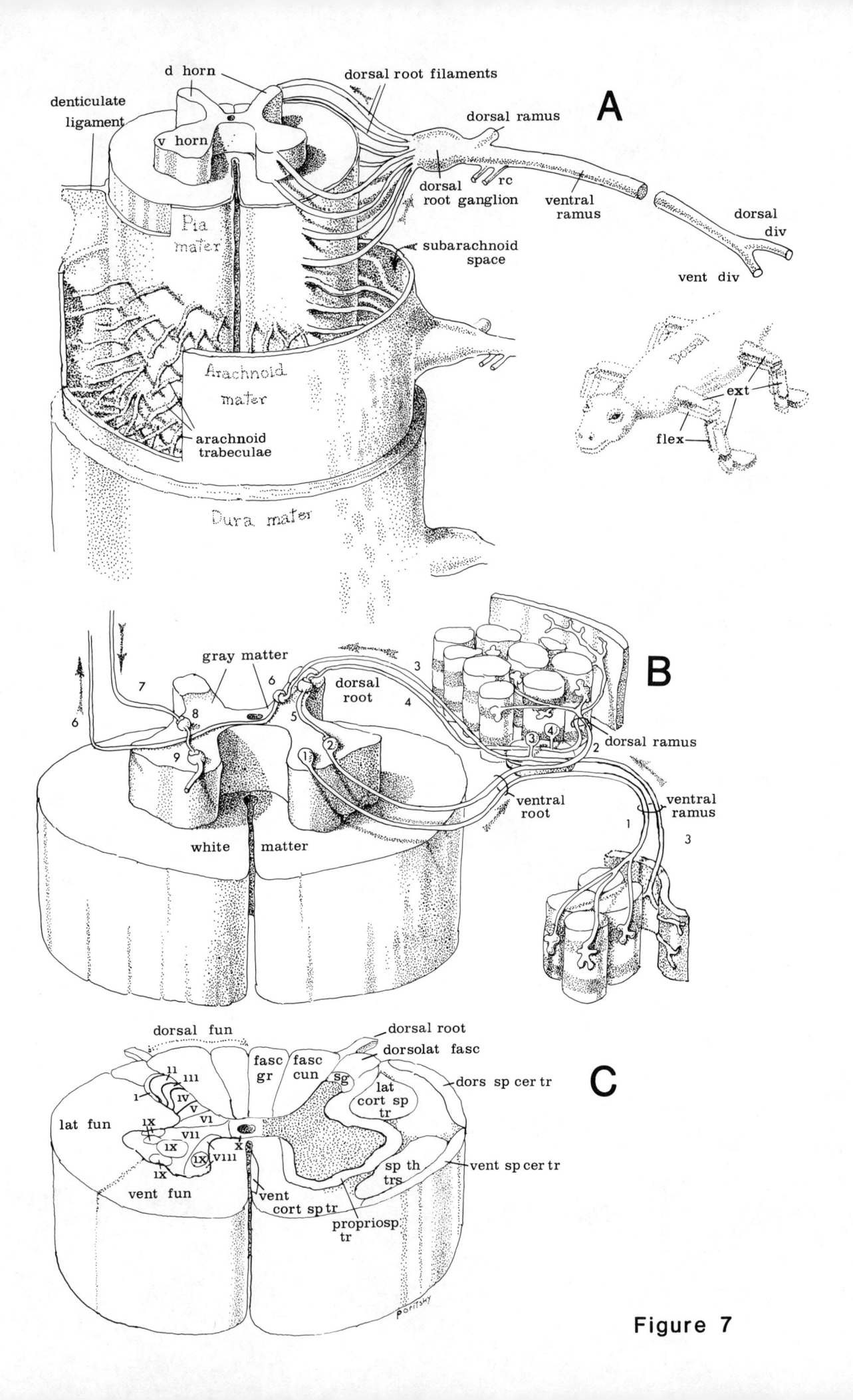

Figure 8

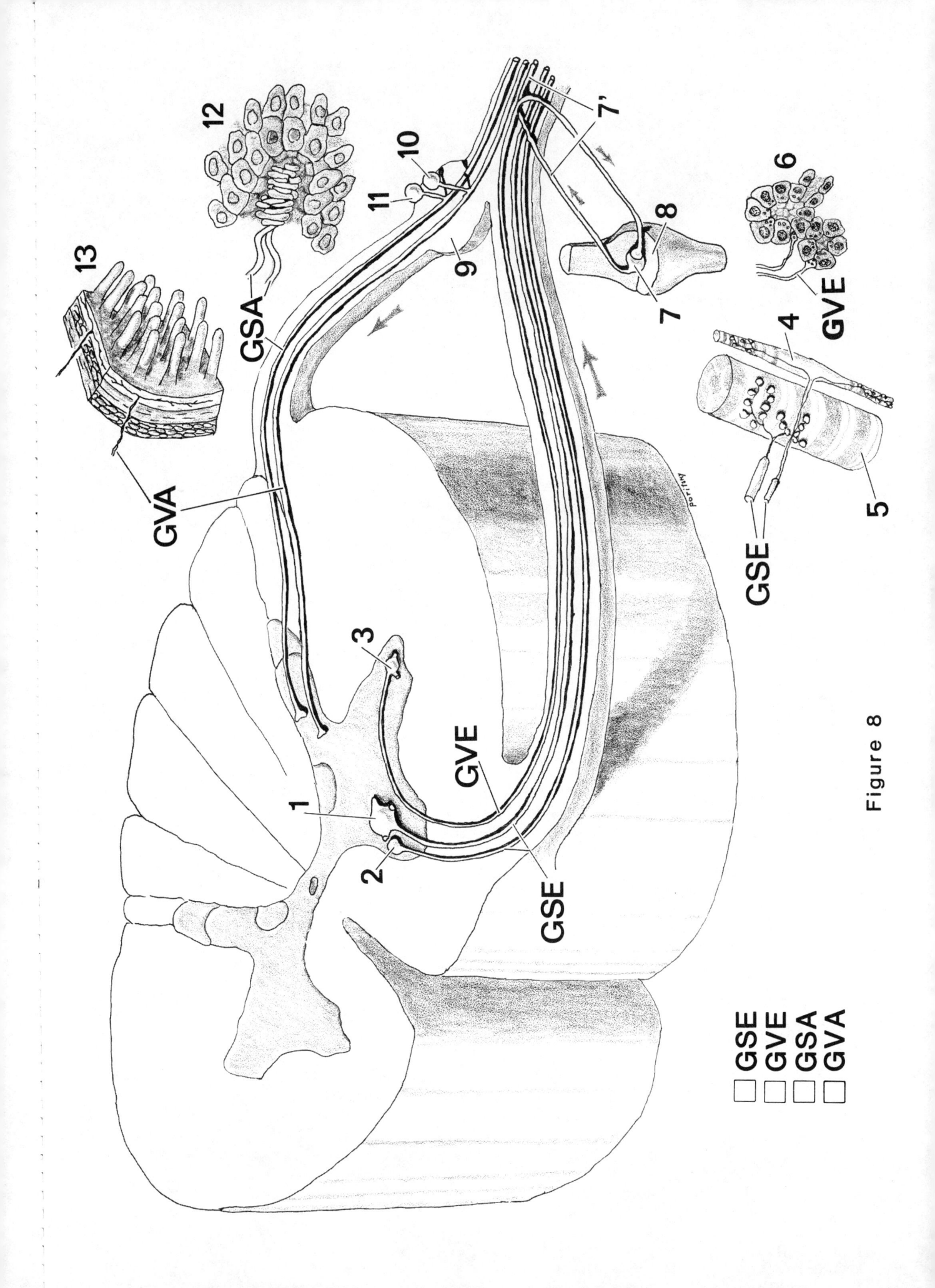

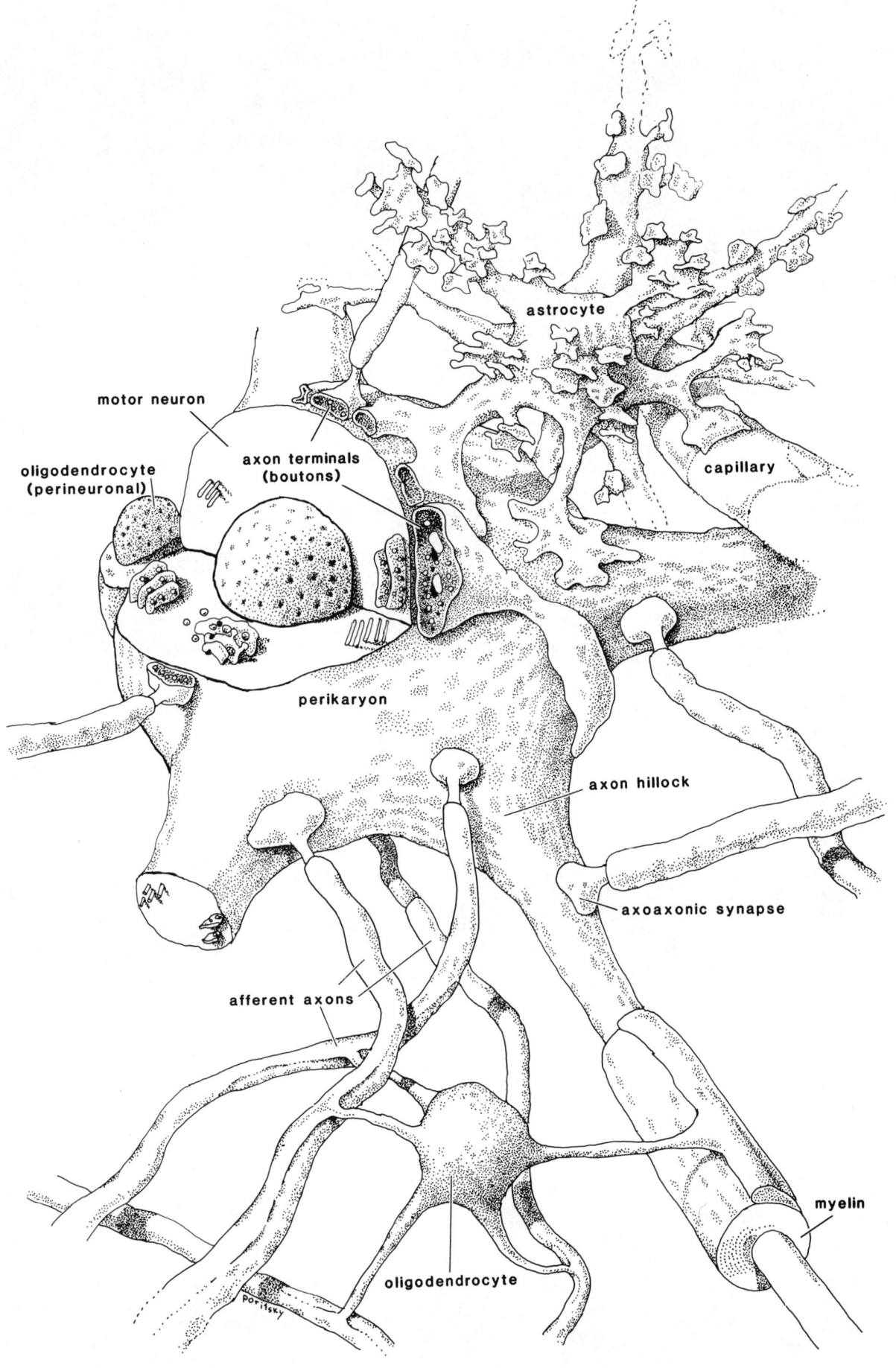

Figure 9

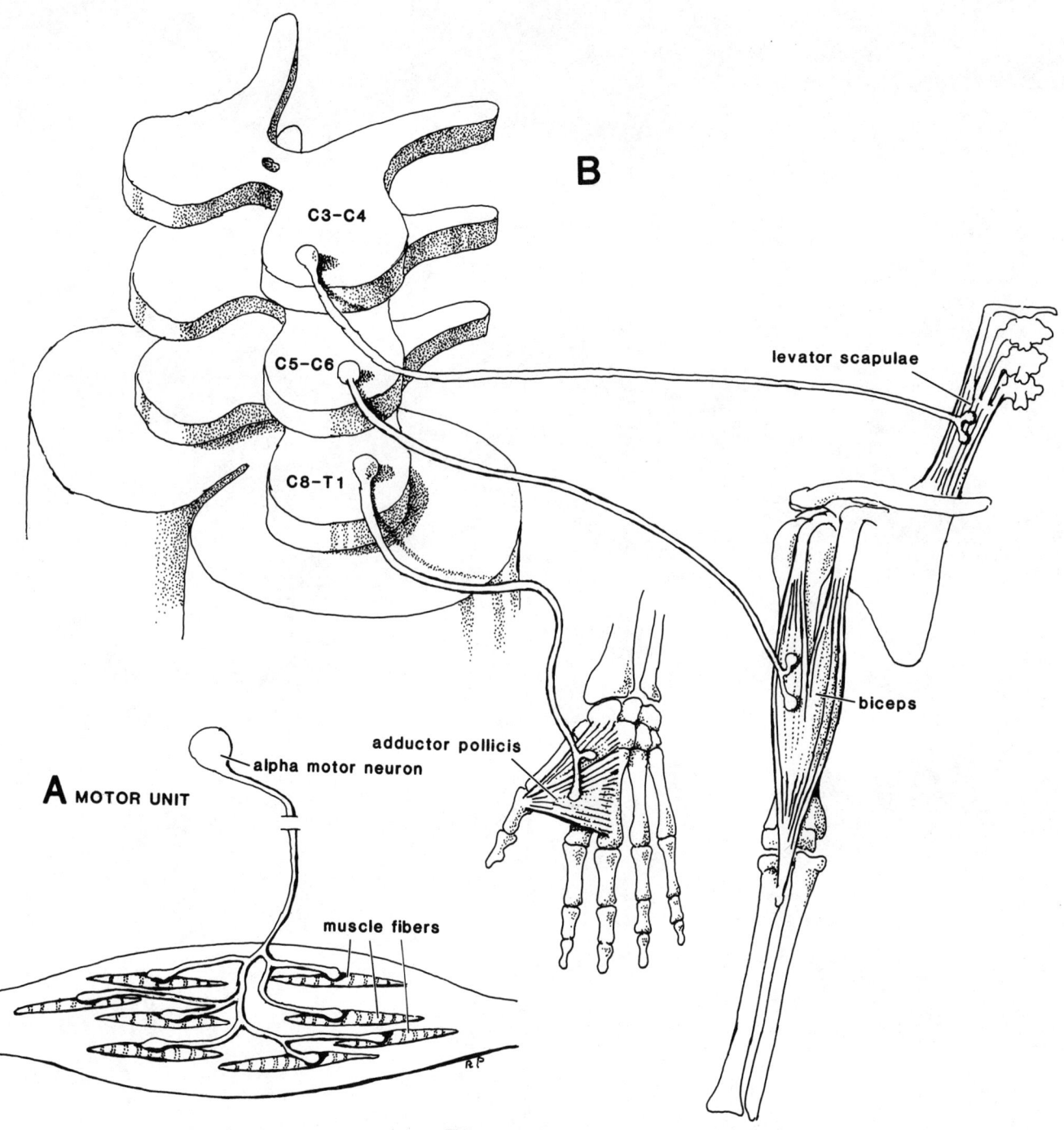

Figure 10

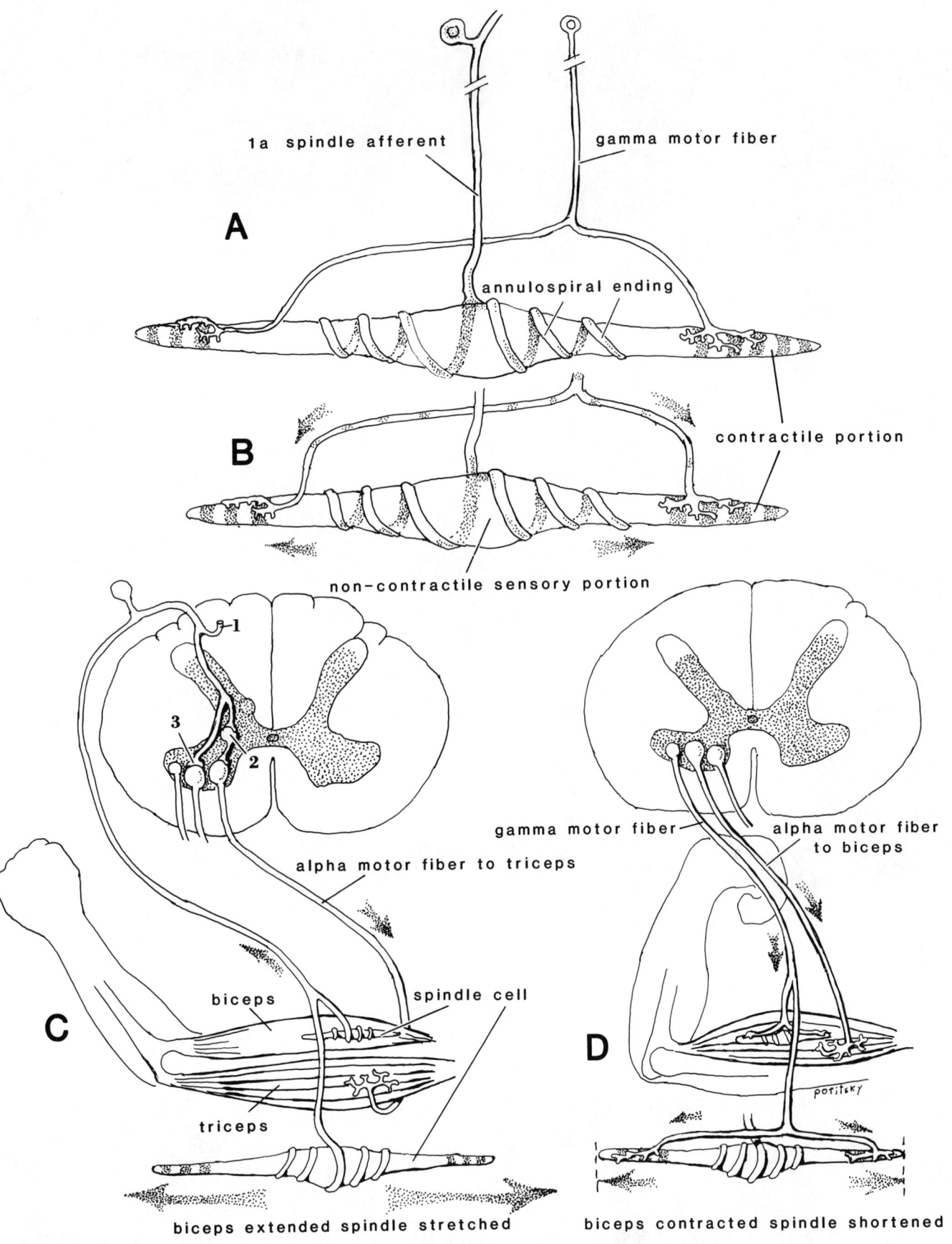

Figure 11

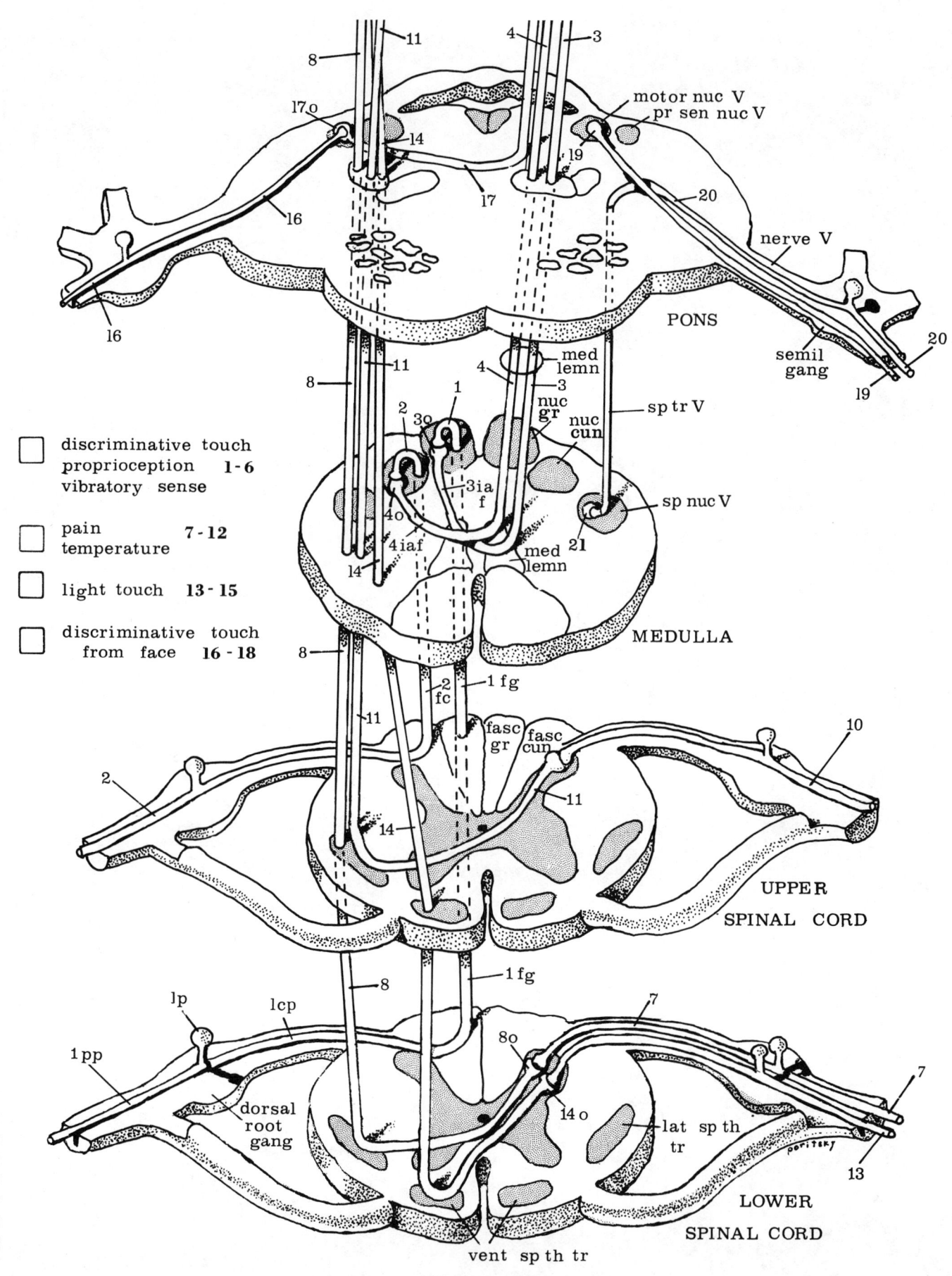

Figure 12

Figure 13

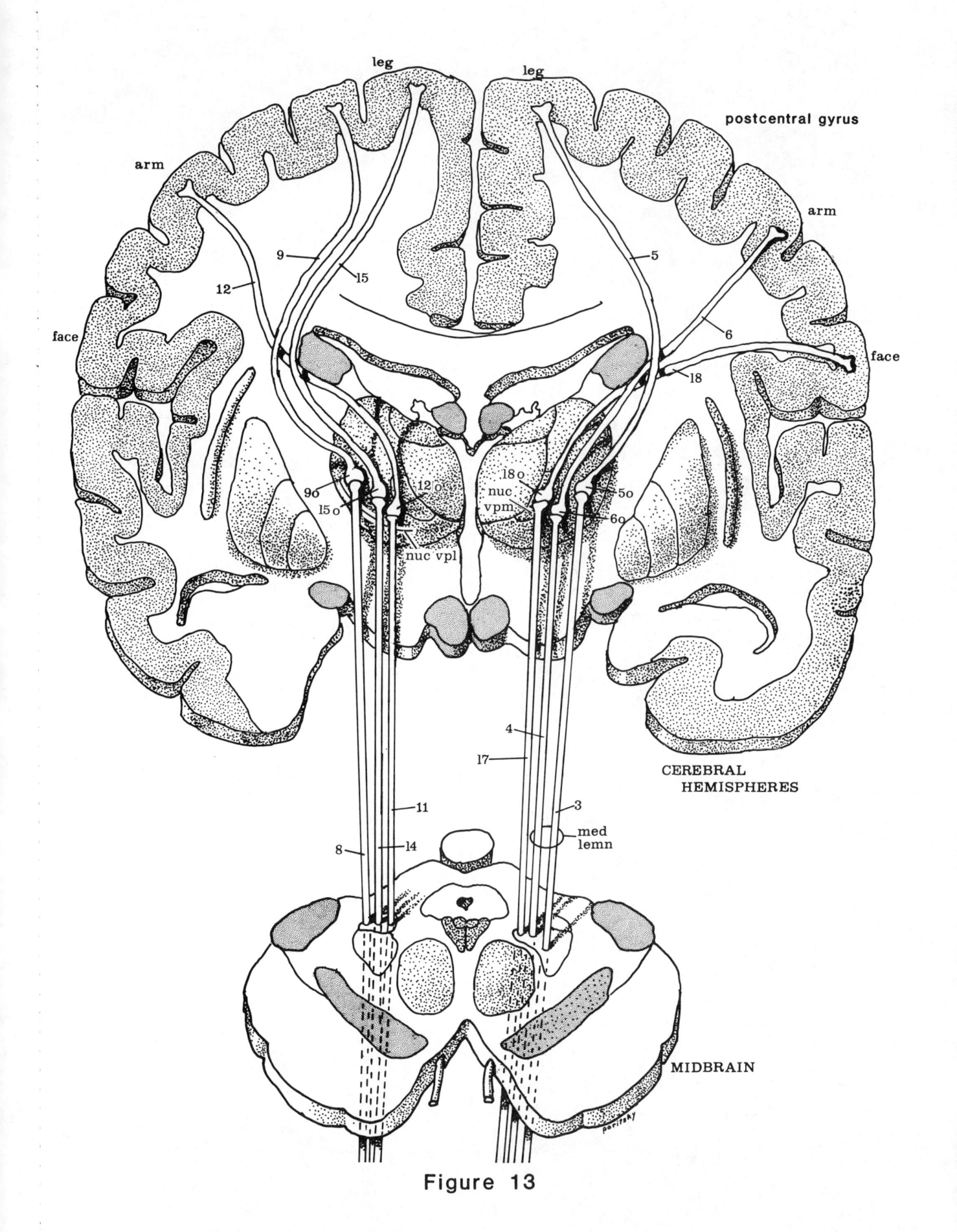

Figure 14

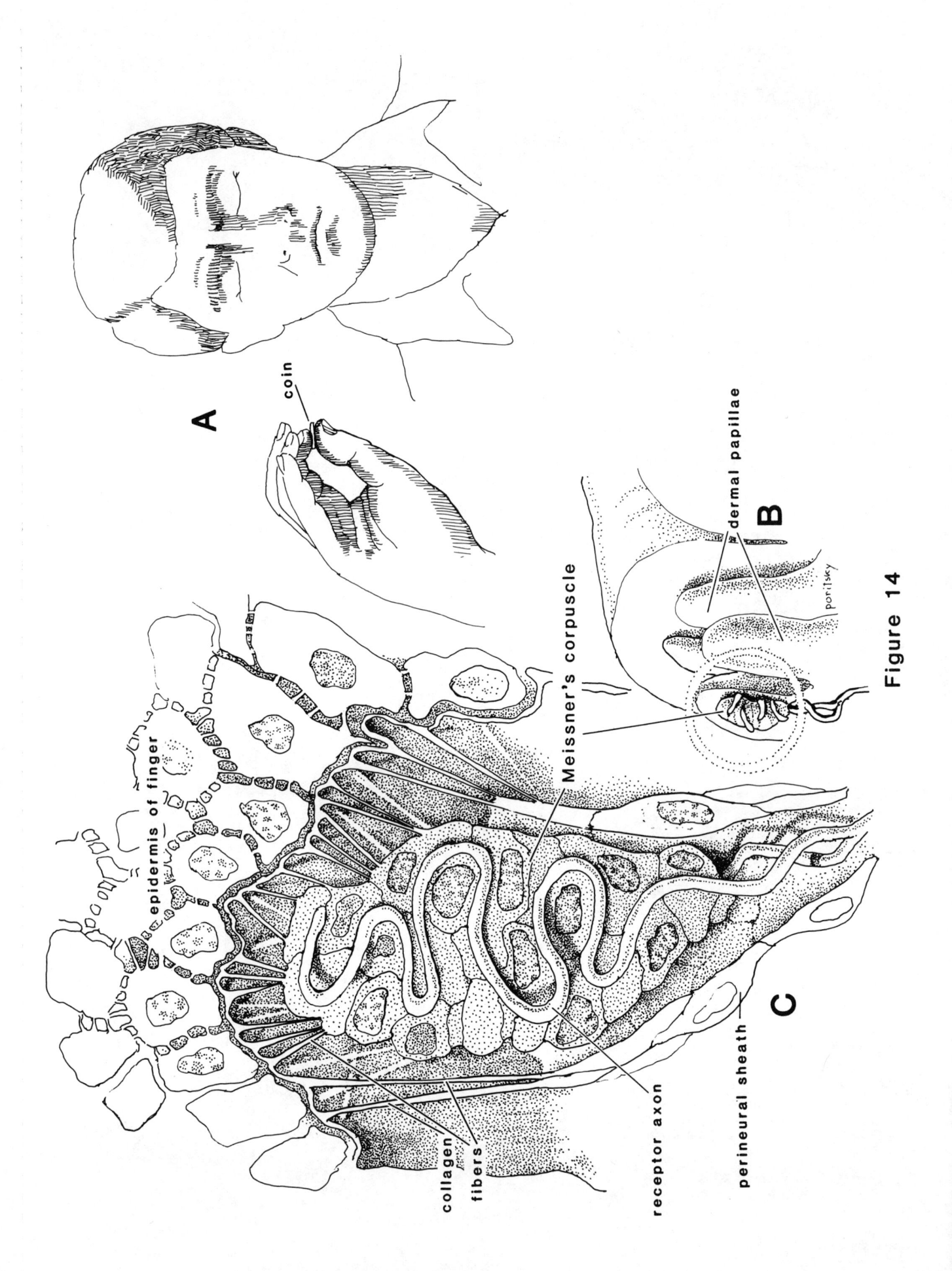

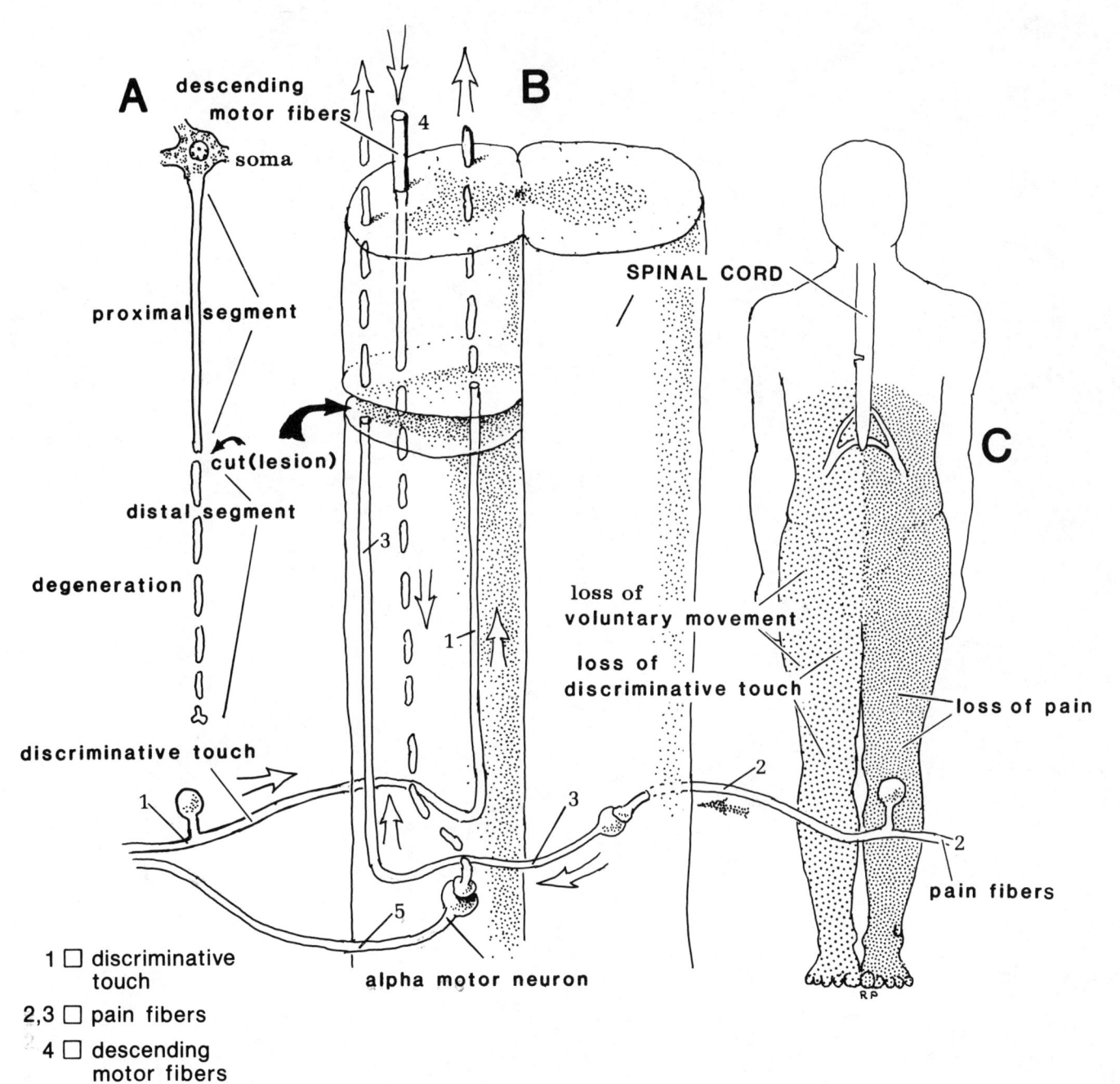

Figure 15

Figure 16

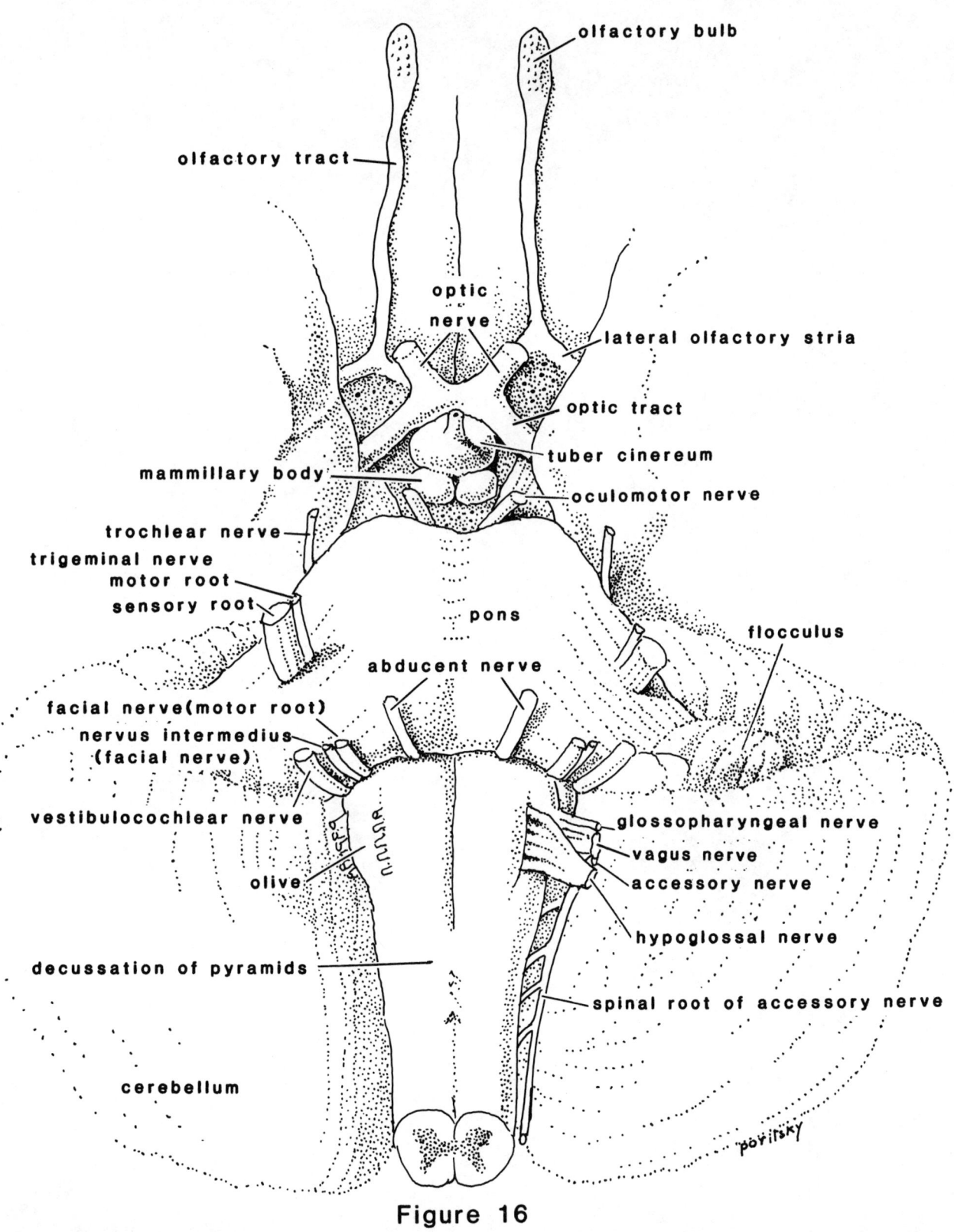

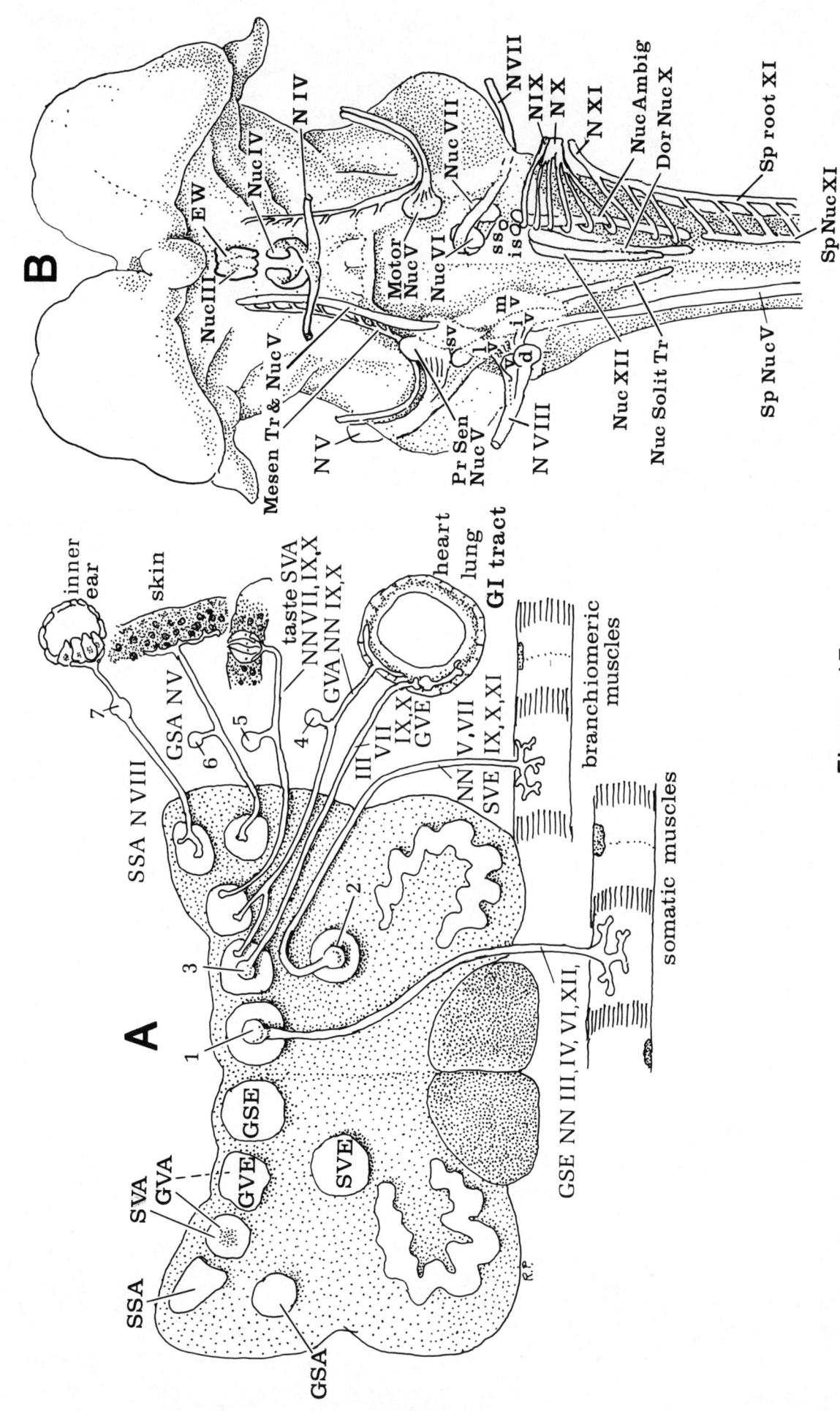

Figure 17

Figure 18

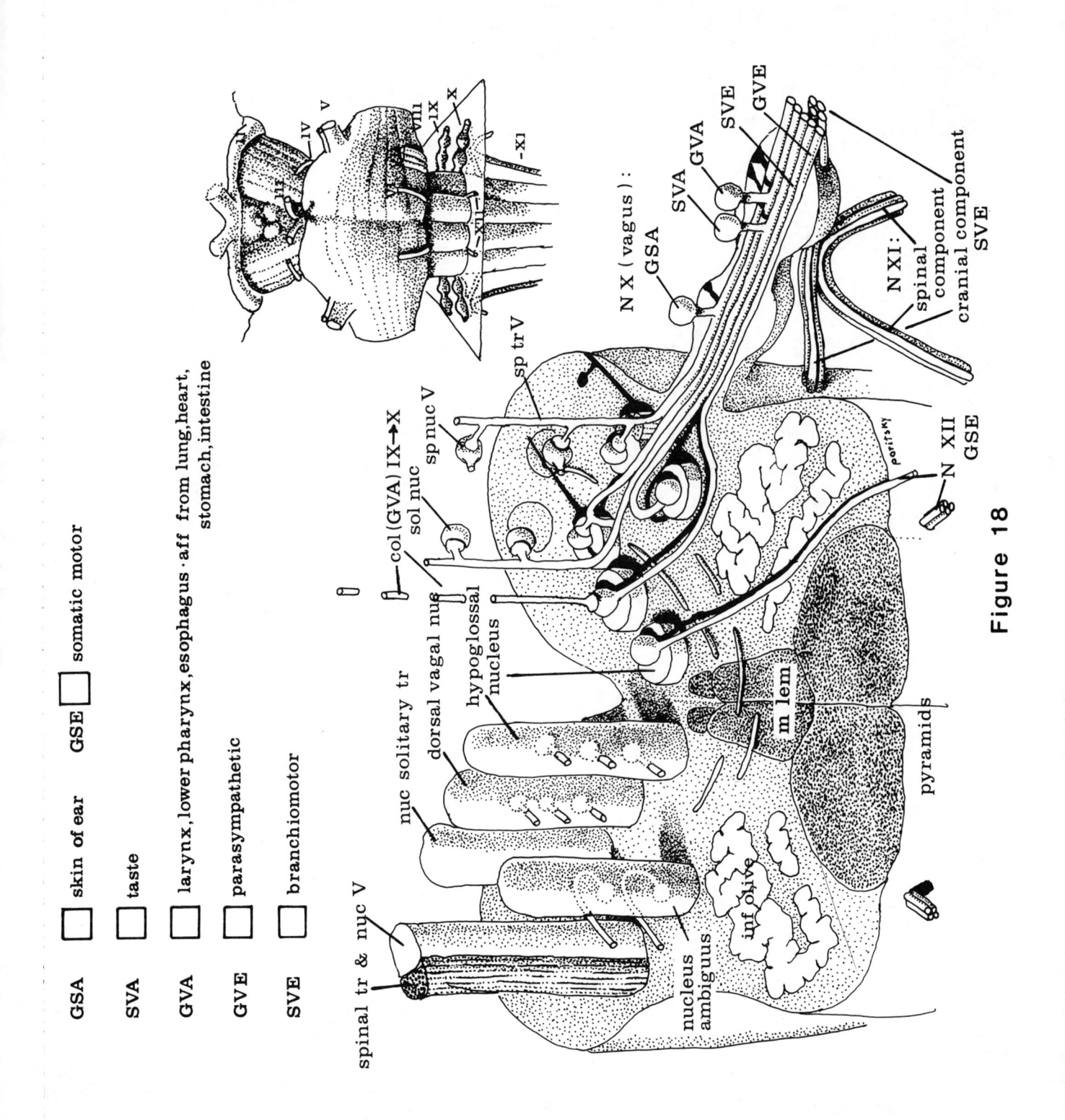

Figure 19

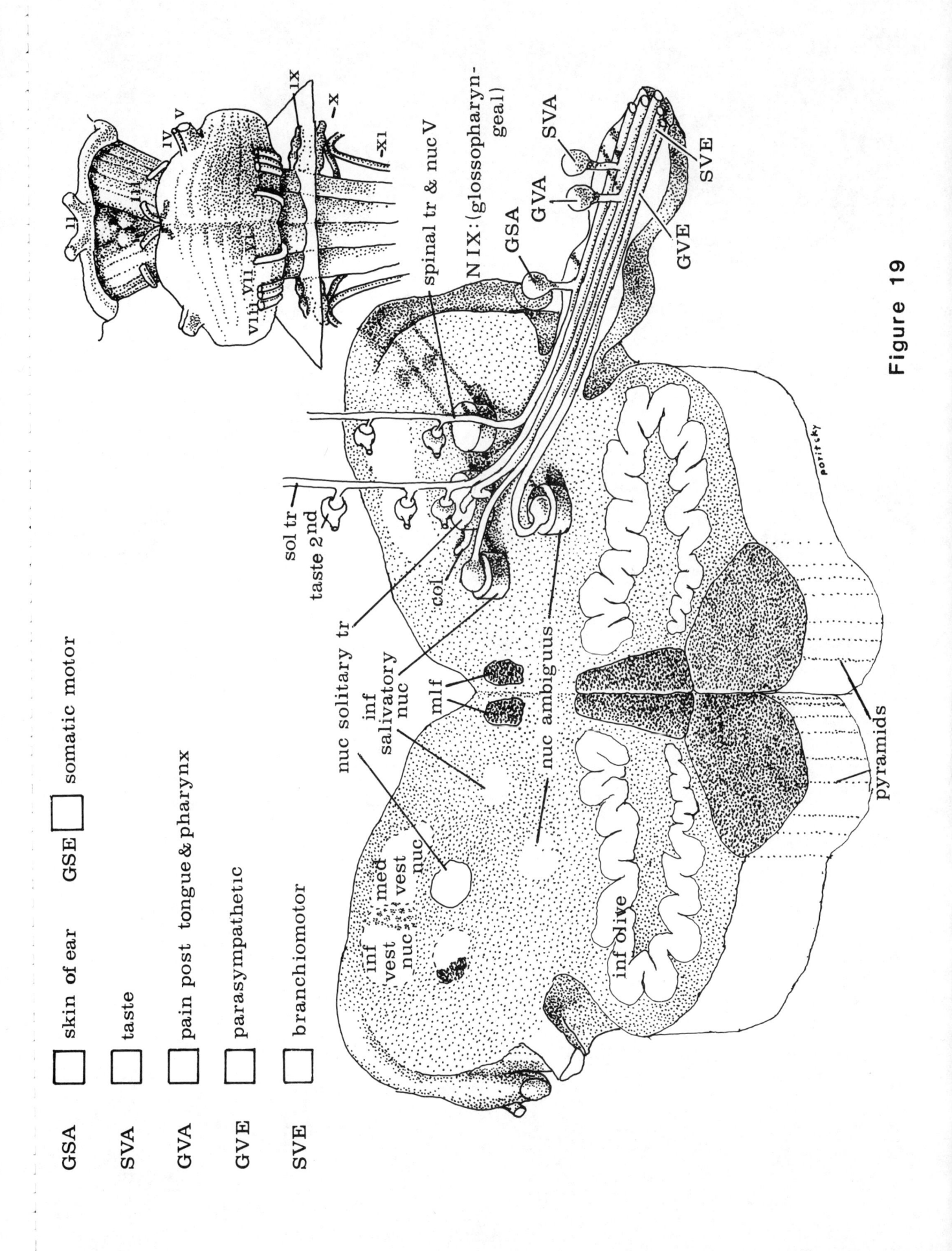

Figure 20

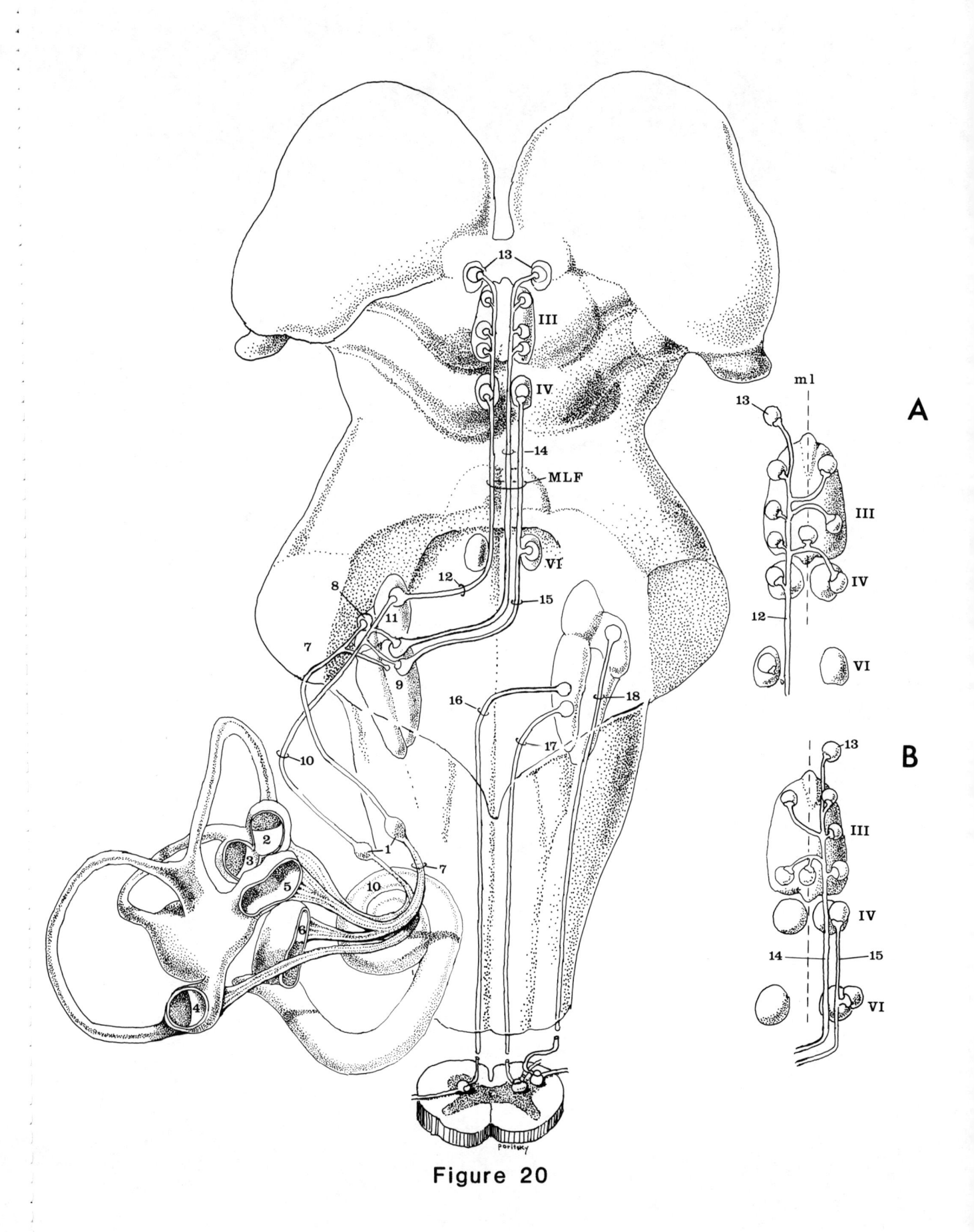

Figure 21

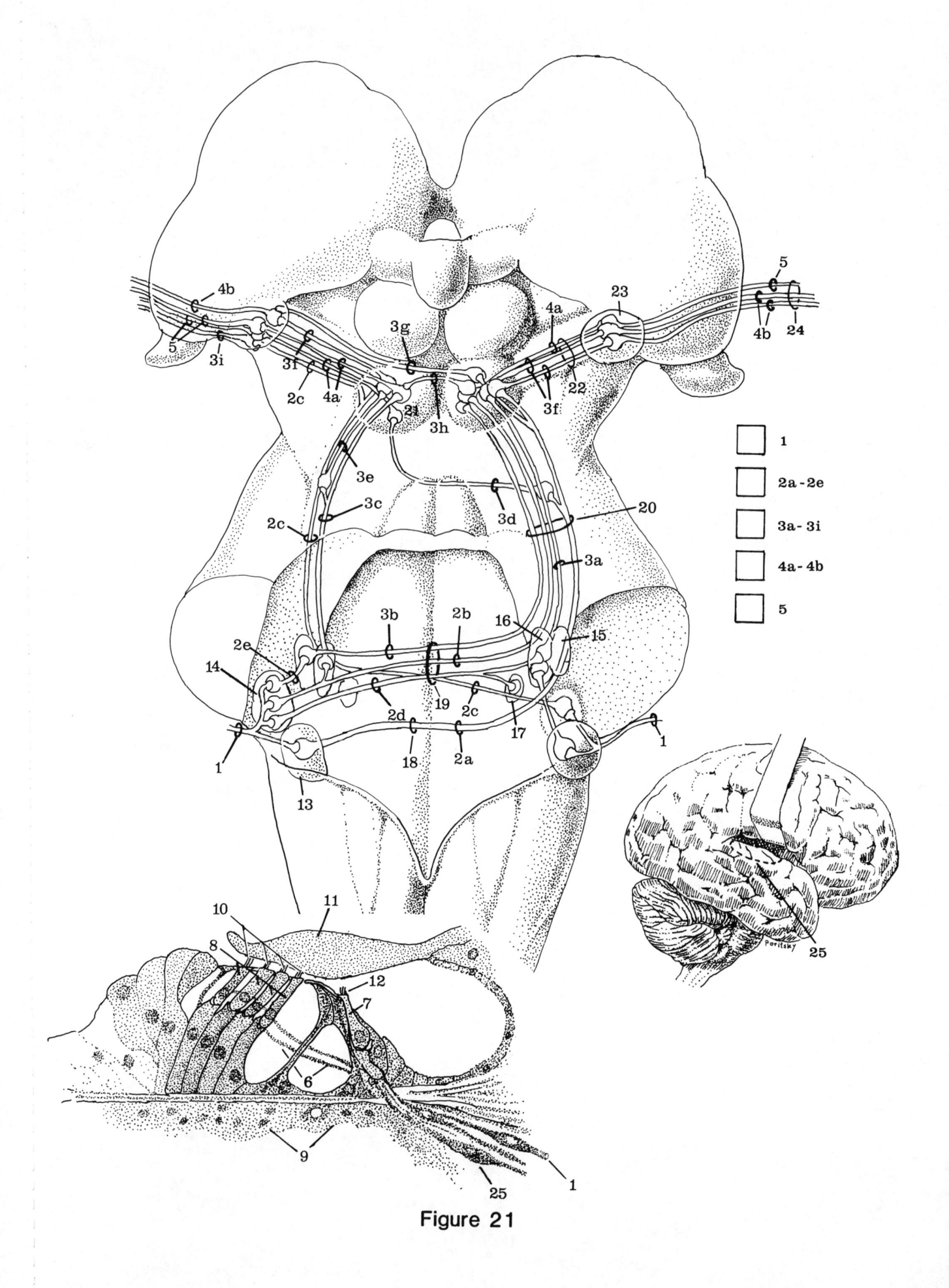

Figure 22

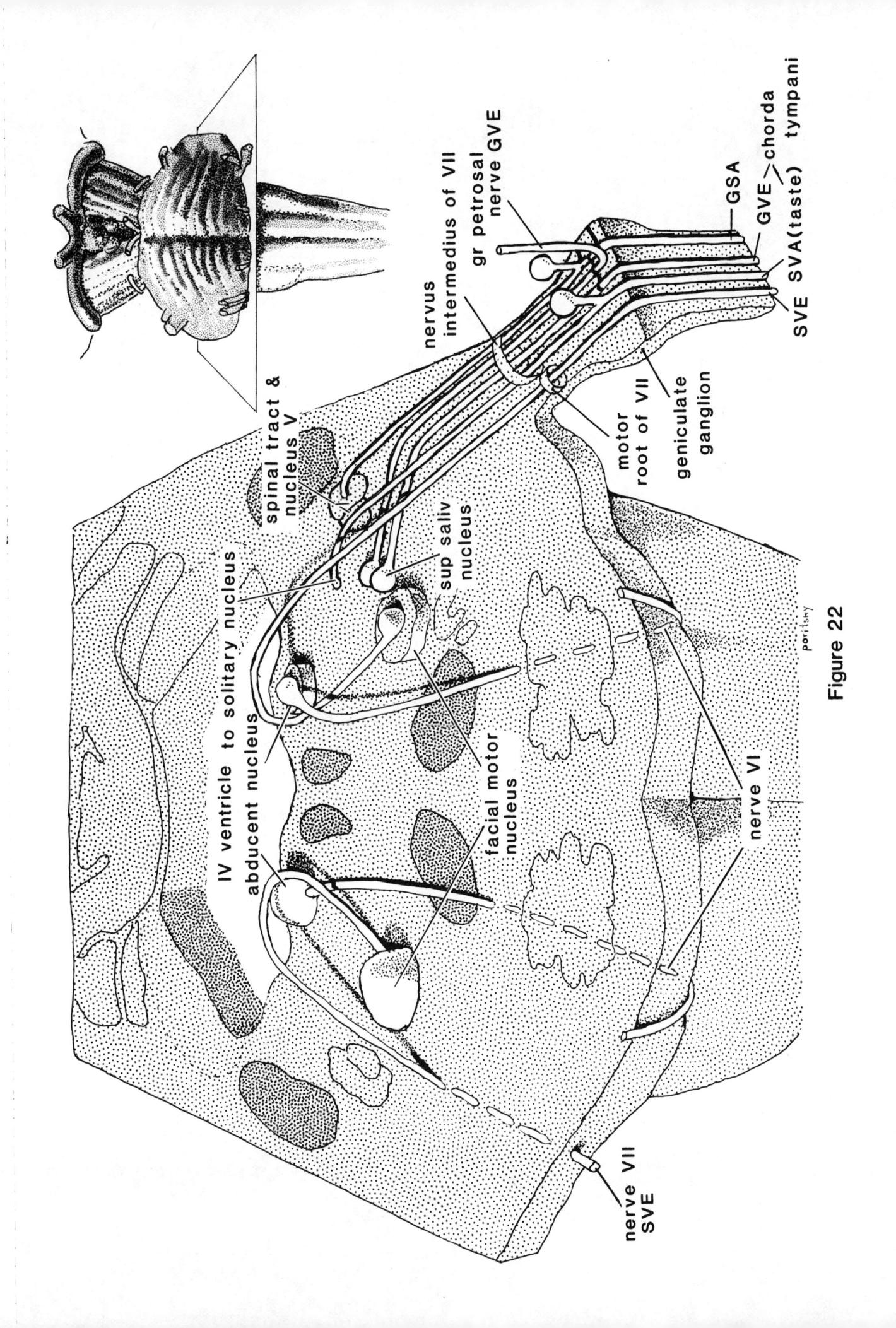

Figure 23

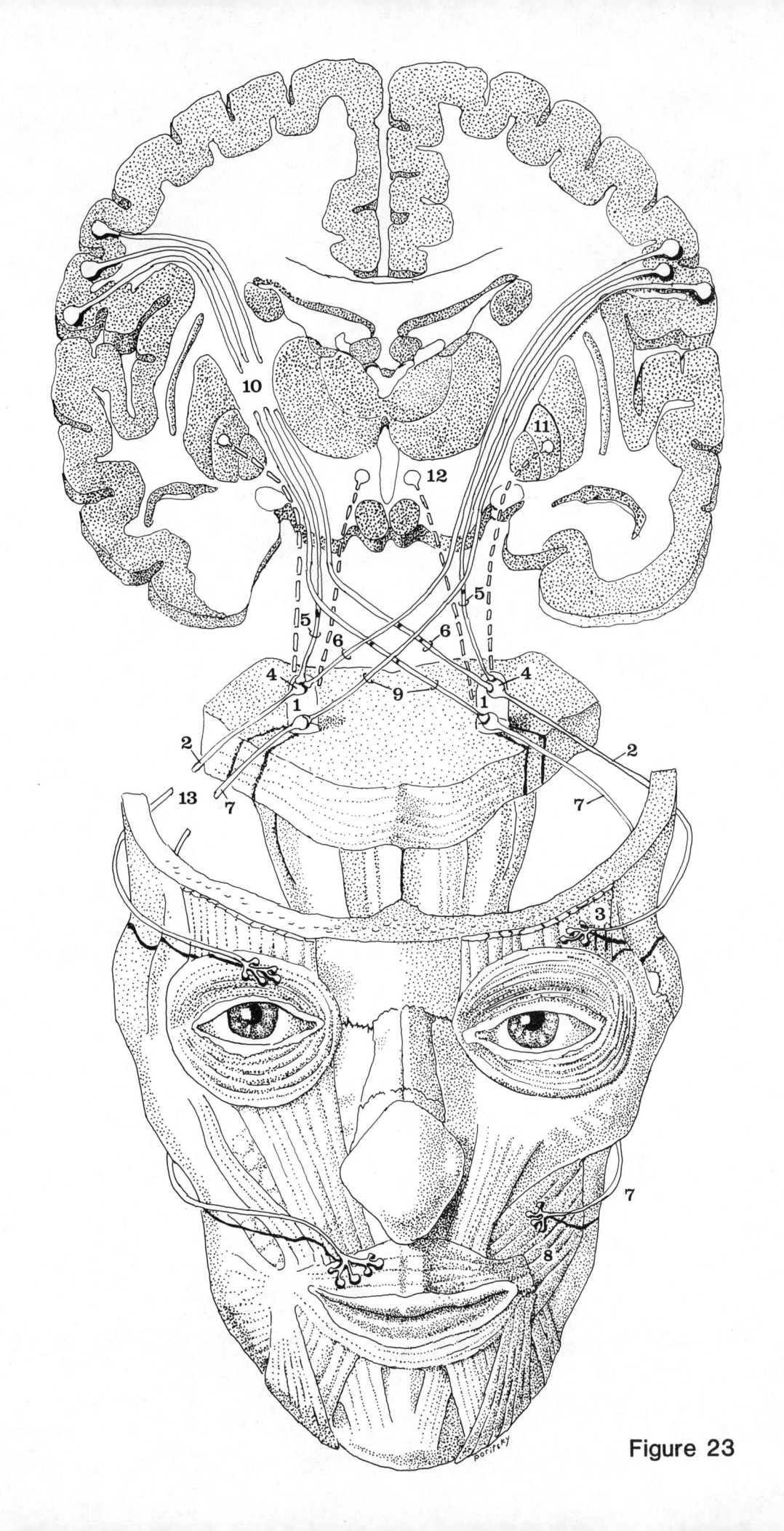

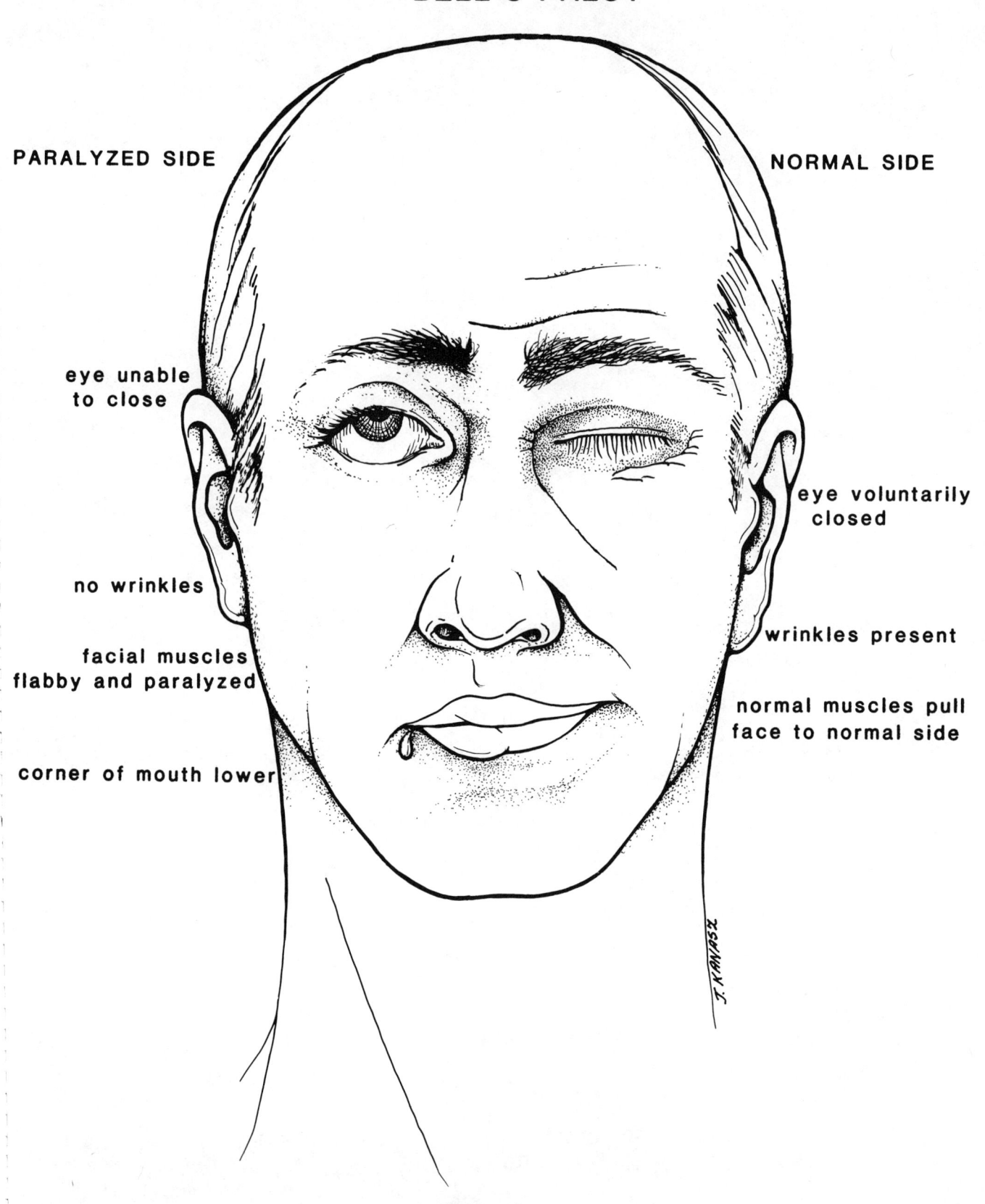

Figure 24

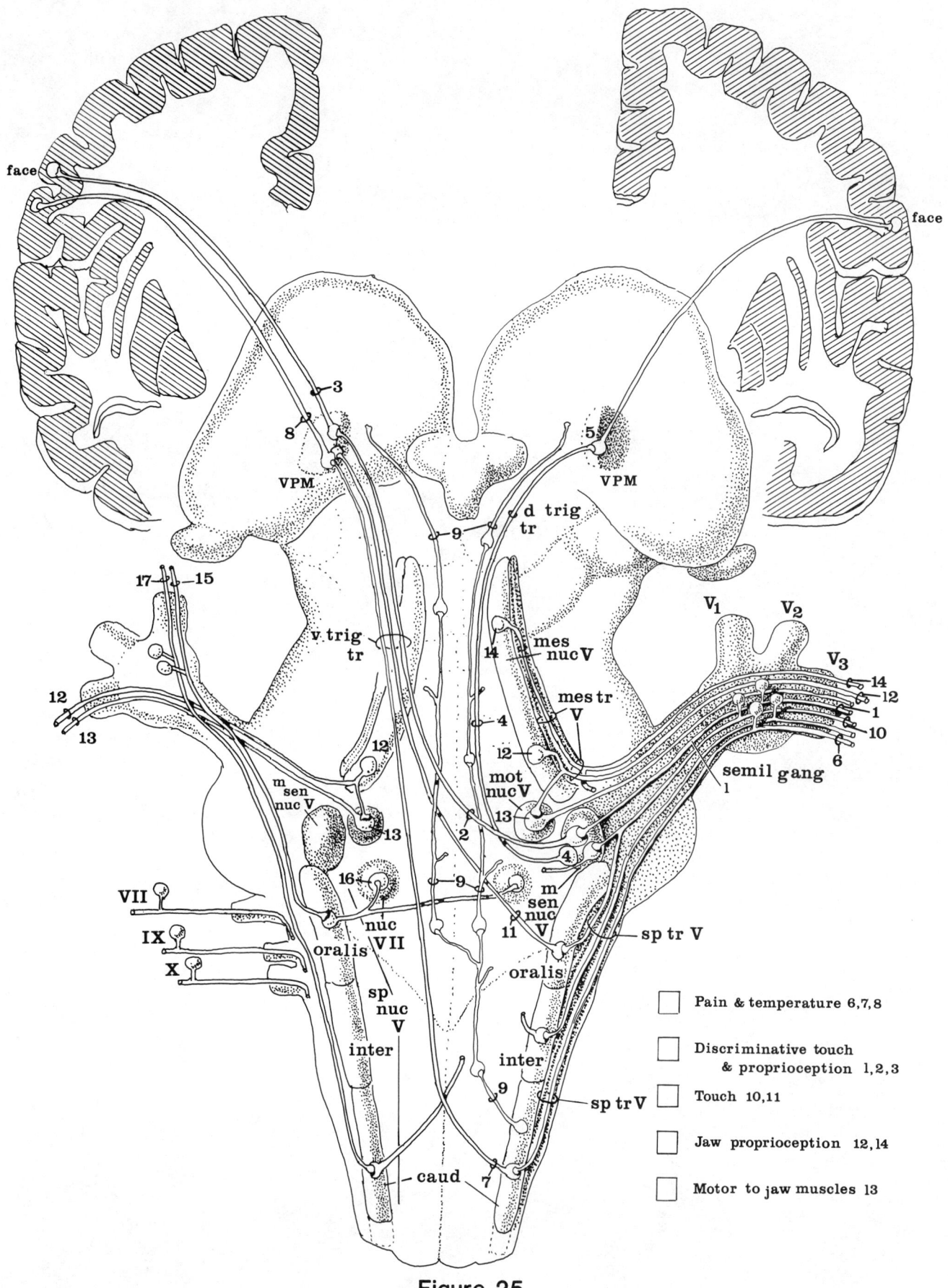

Figure 25

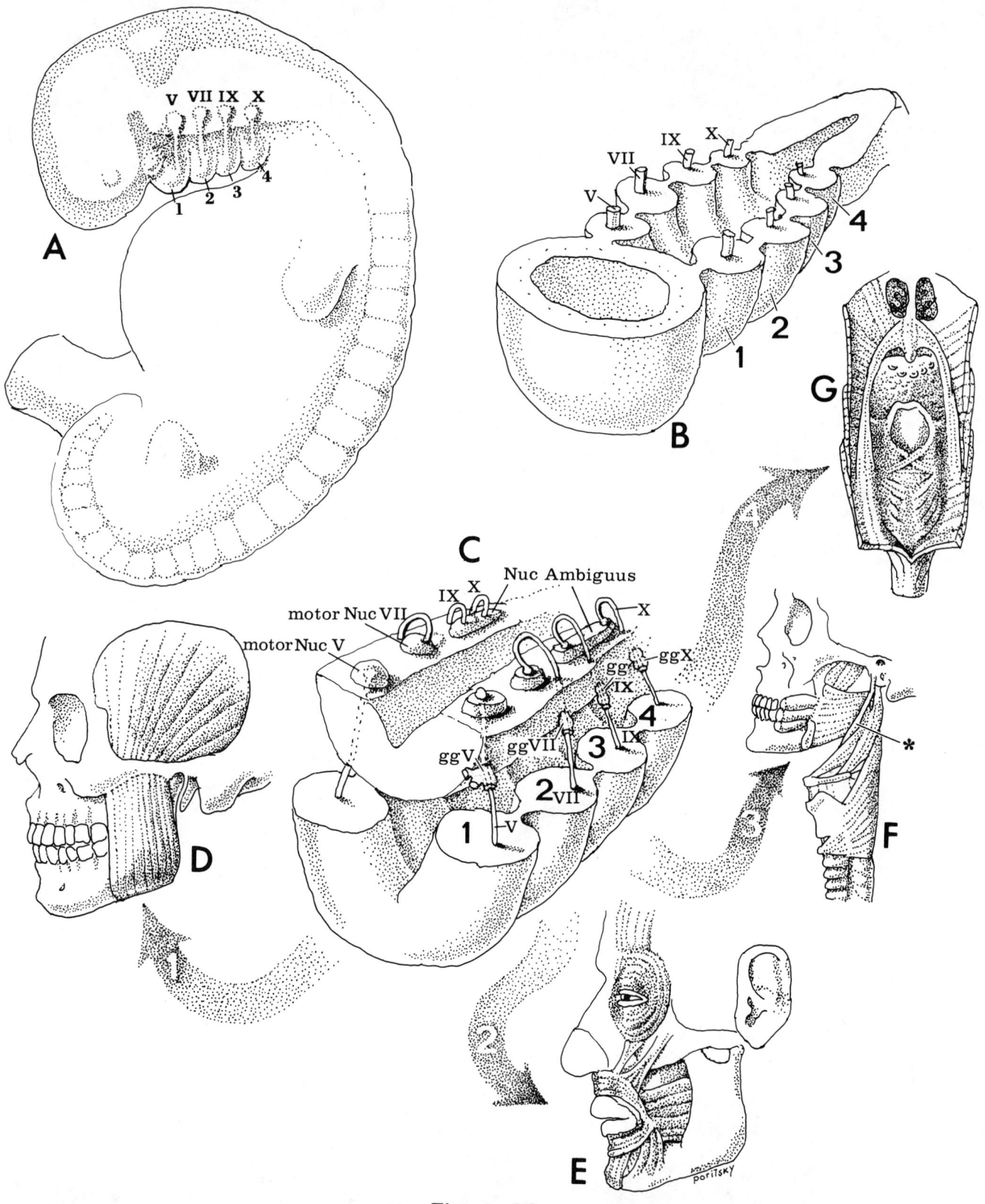

Figure 26

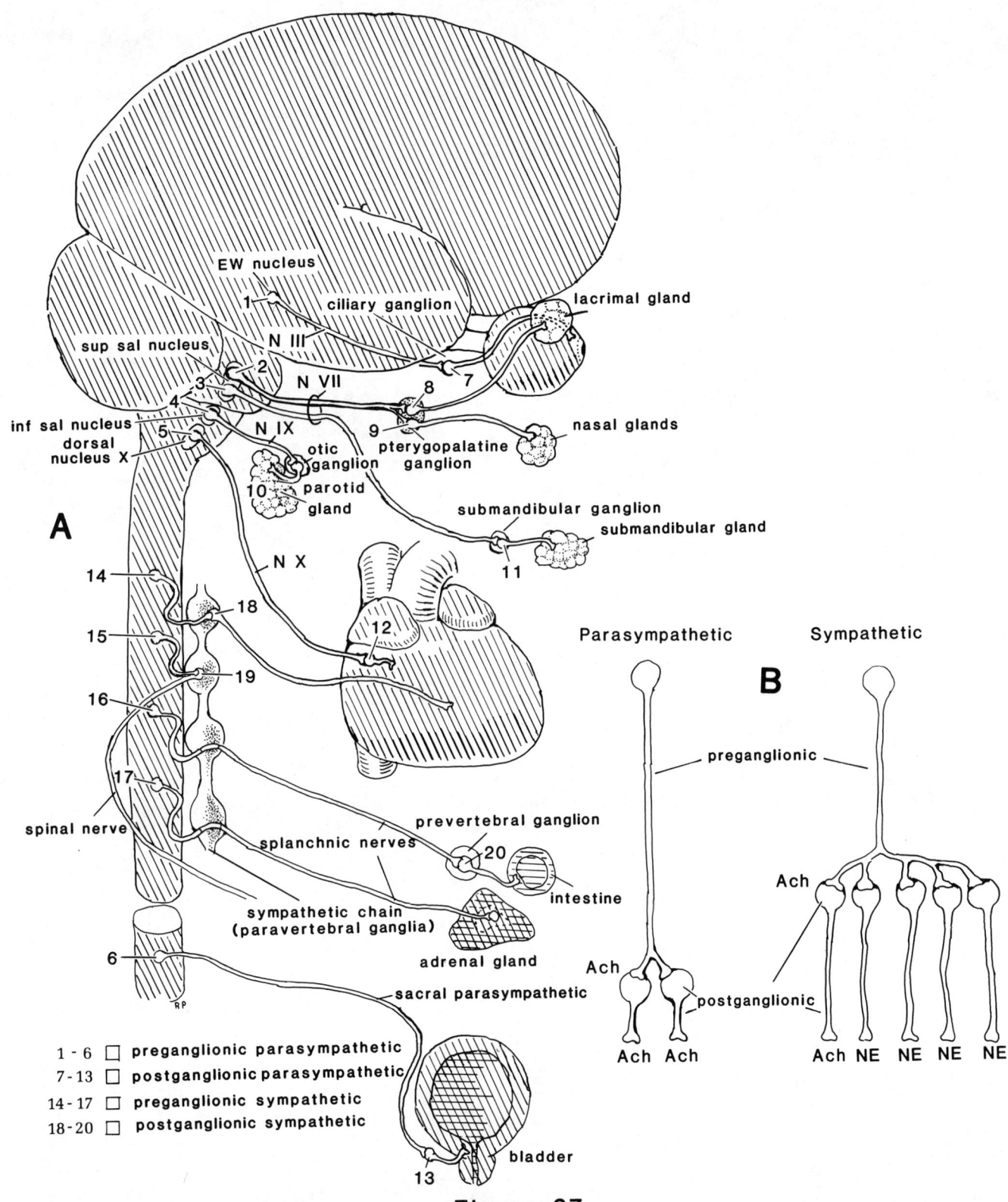

Figure 27

Figure 28

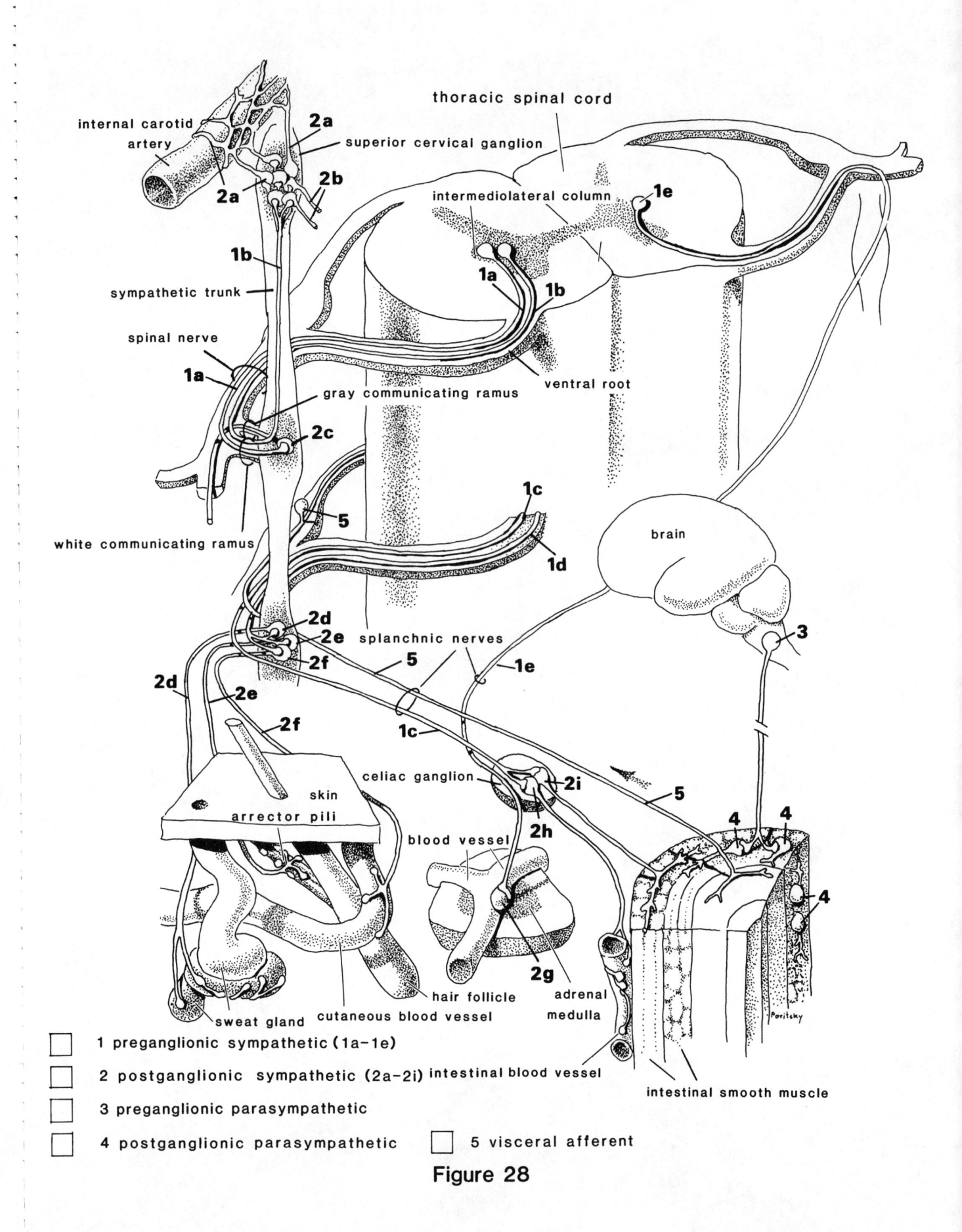

Figure 29

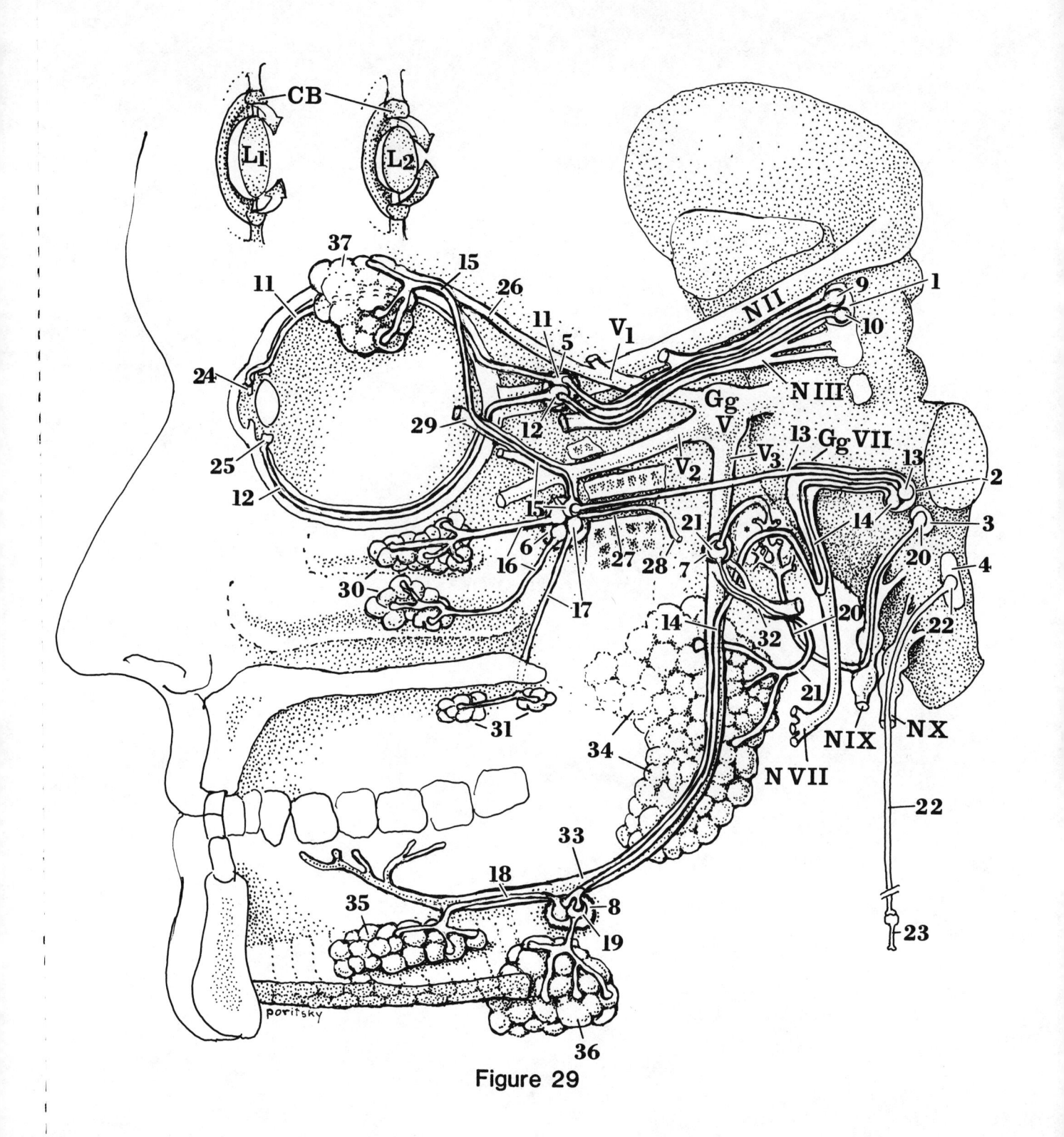

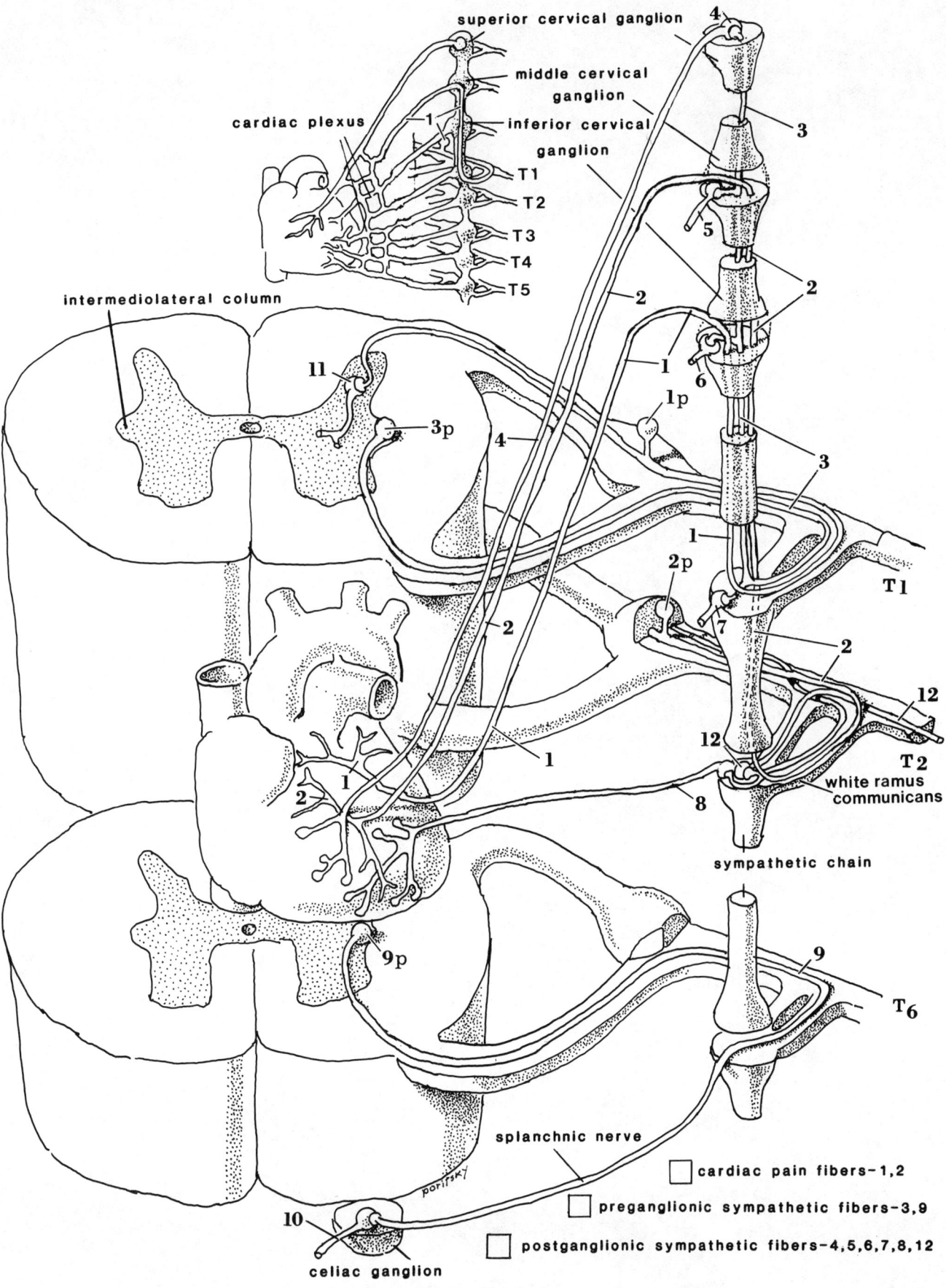

Figure 30

visual field

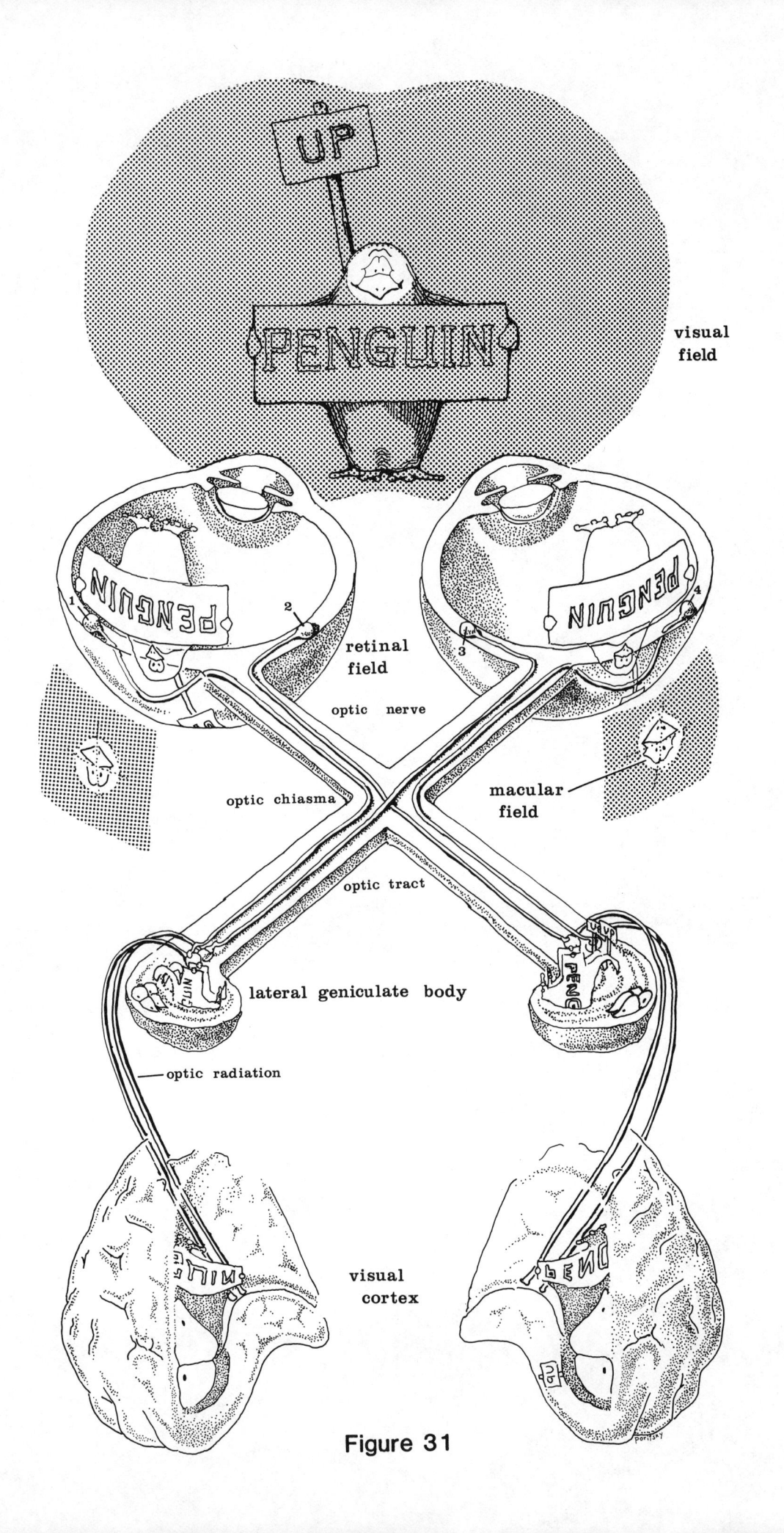

retinal field

optic nerve

optic chiasma

macular field

optic tract

lateral geniculate body

optic radiation

visual cortex

Figure 31

Figure 32

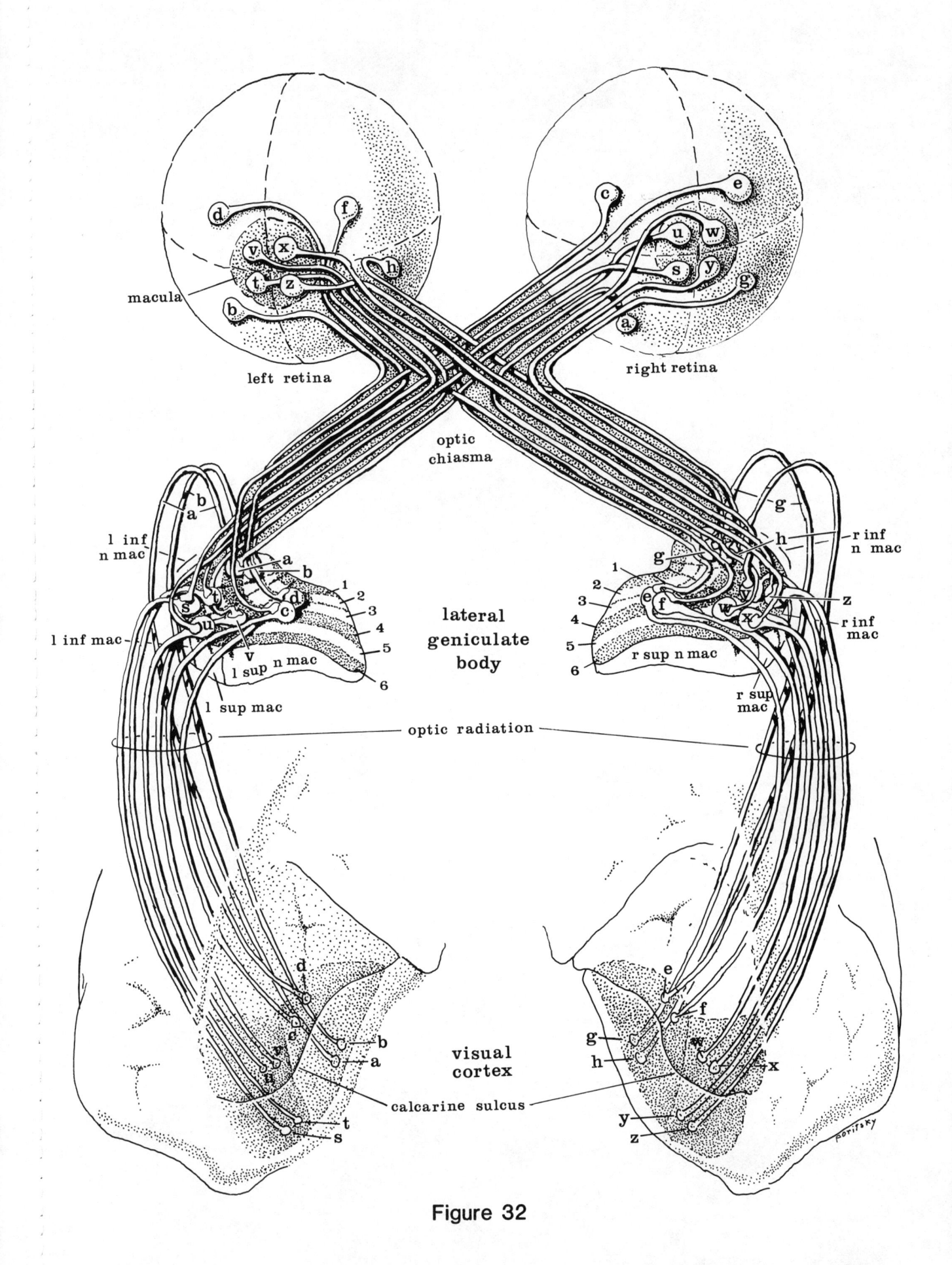

Figure 33

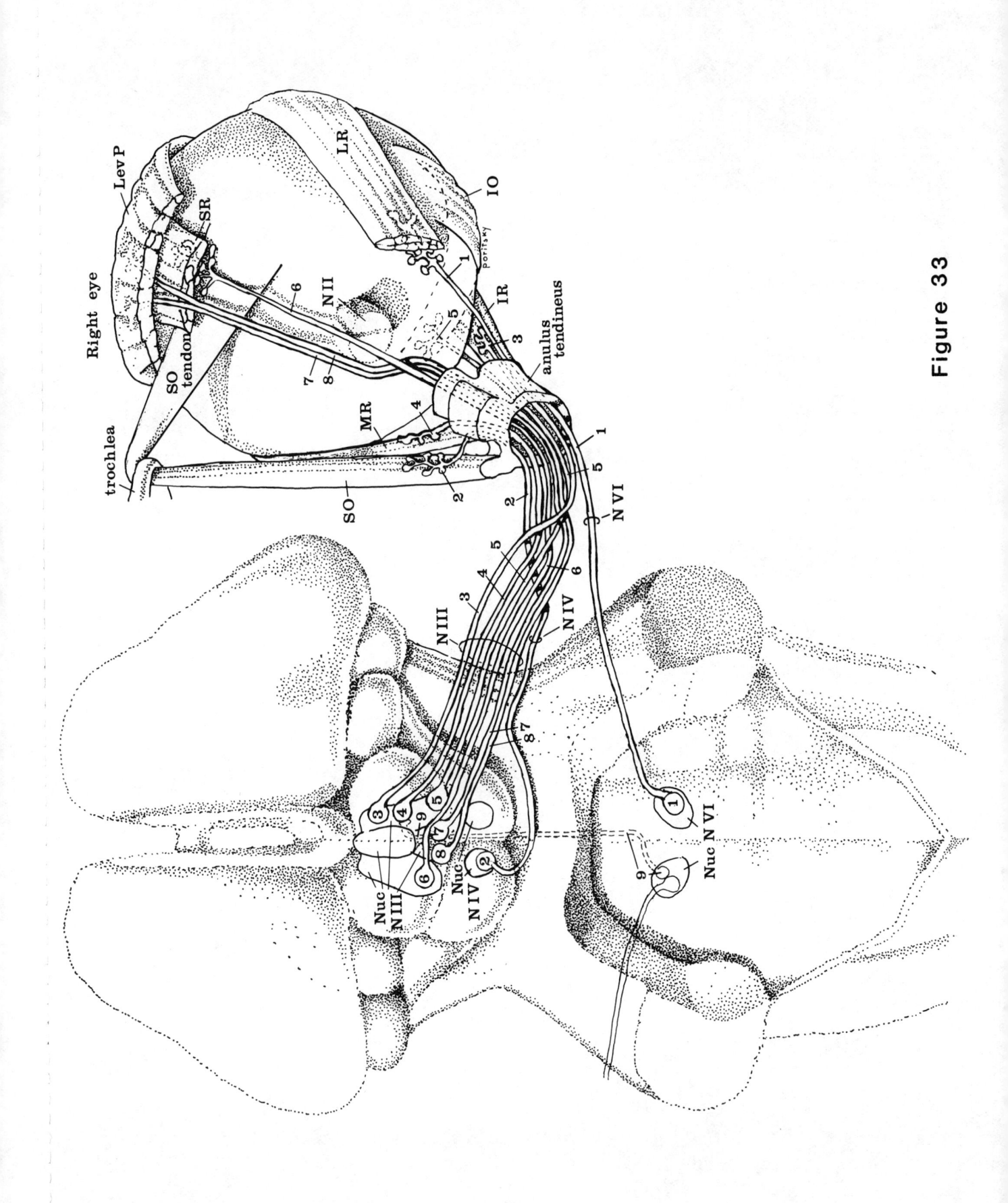

Figure 34

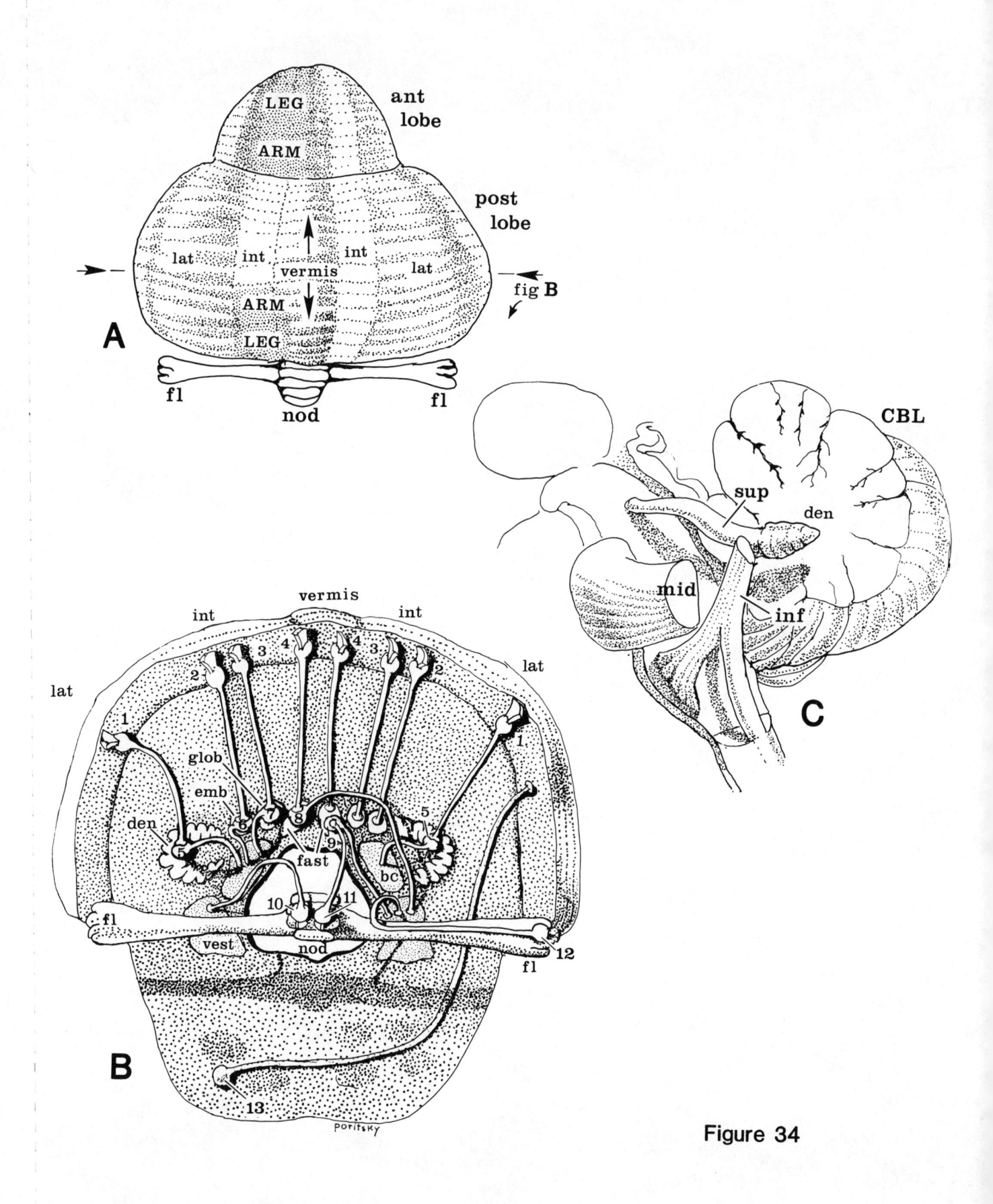

Figure 35

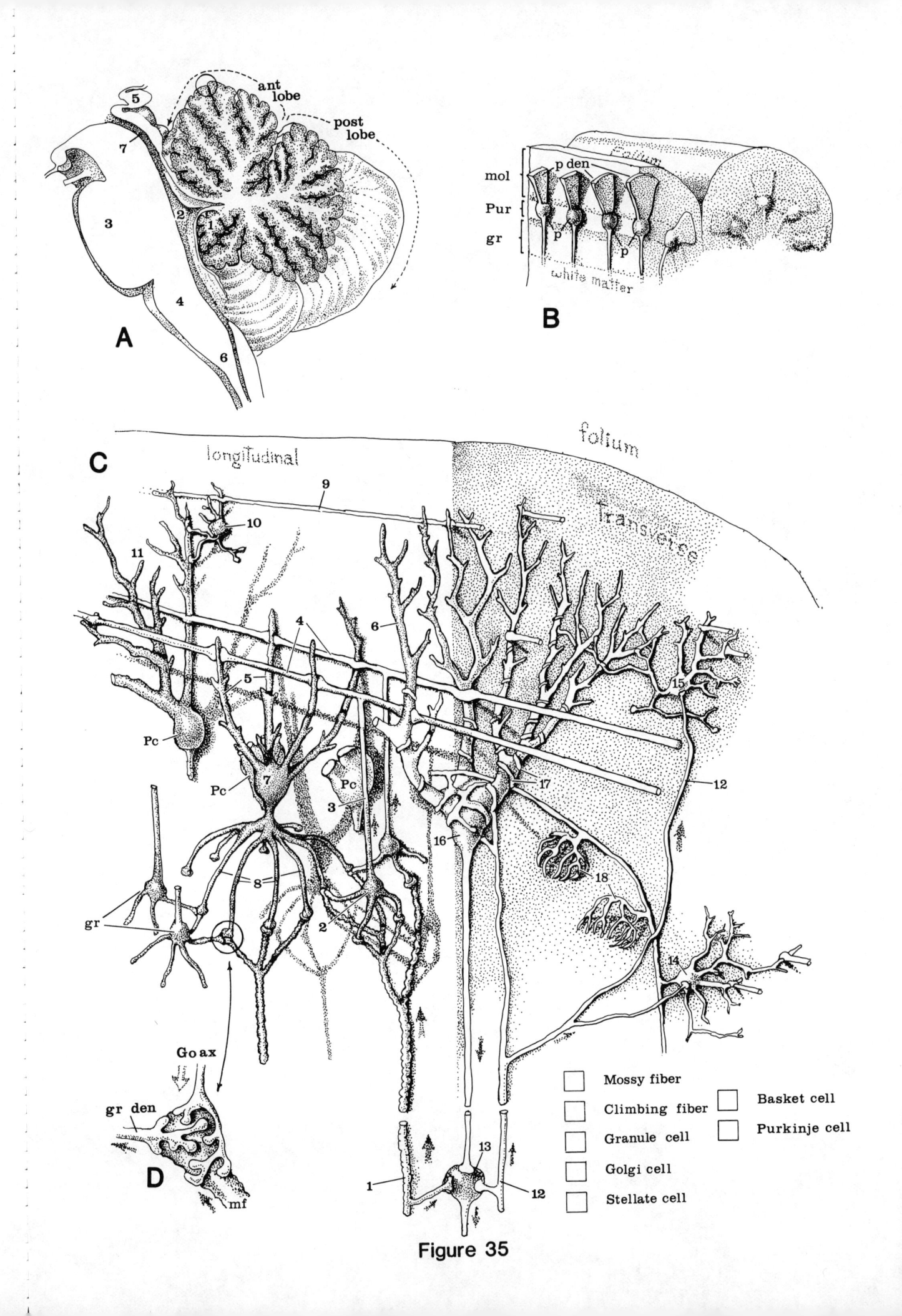

Figure 36

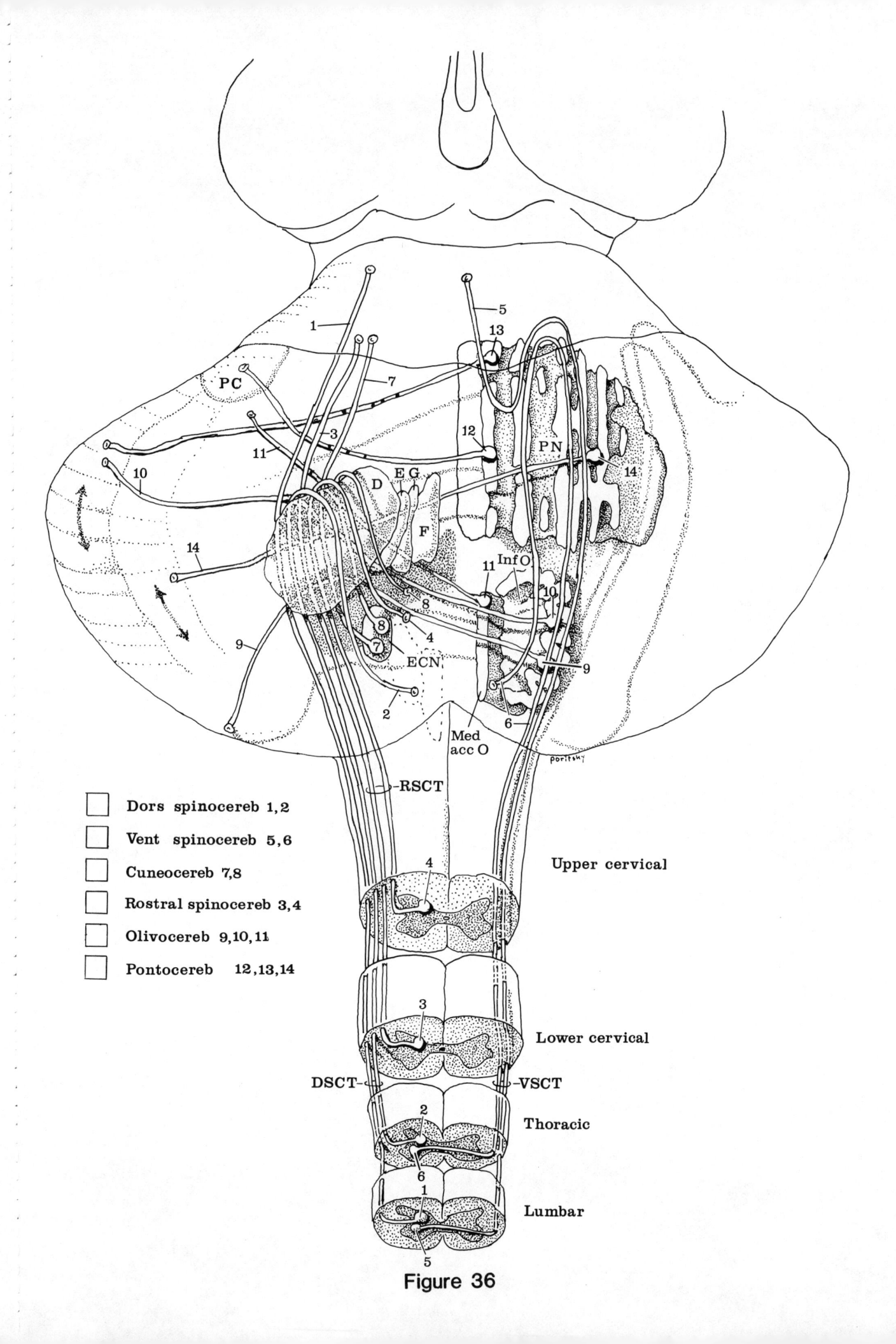

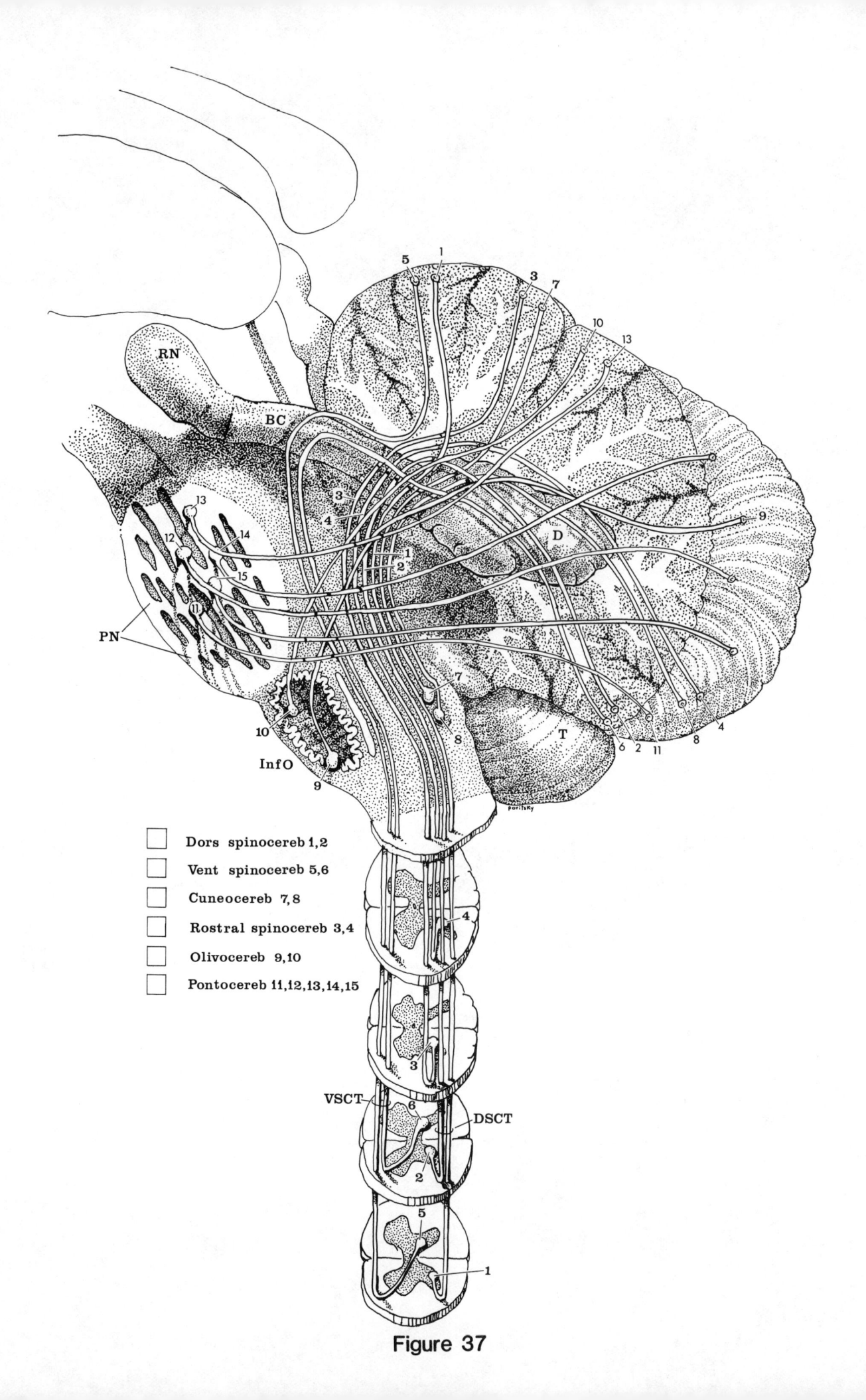

Dors spinocereb 1,2
Vent spinocereb 5,6
Cuneocereb 7,8
Rostral spinocereb 3,4
Olivocereb 9,10
Pontocereb 11,12,13,14,15

Figure 37

Figure 38

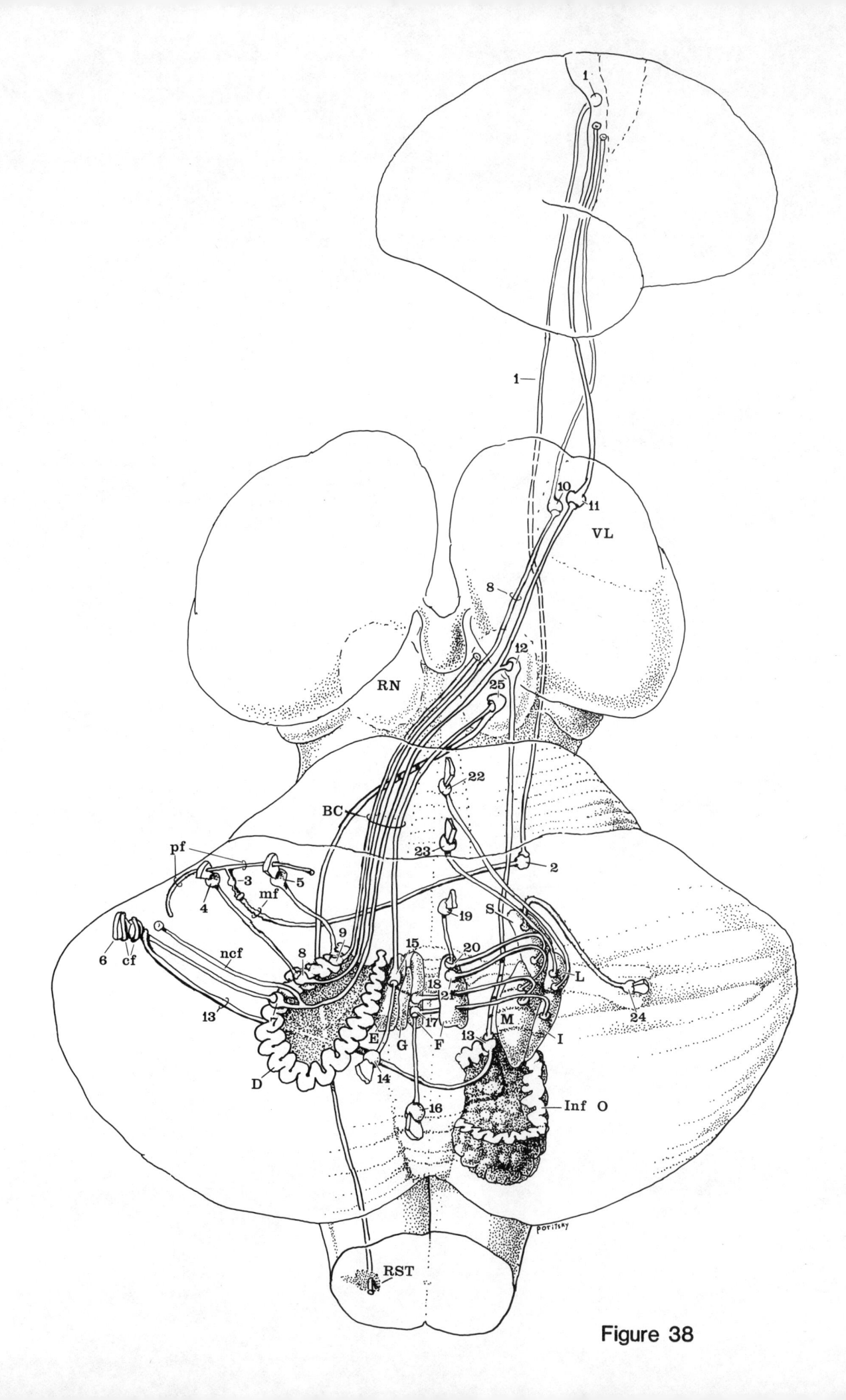

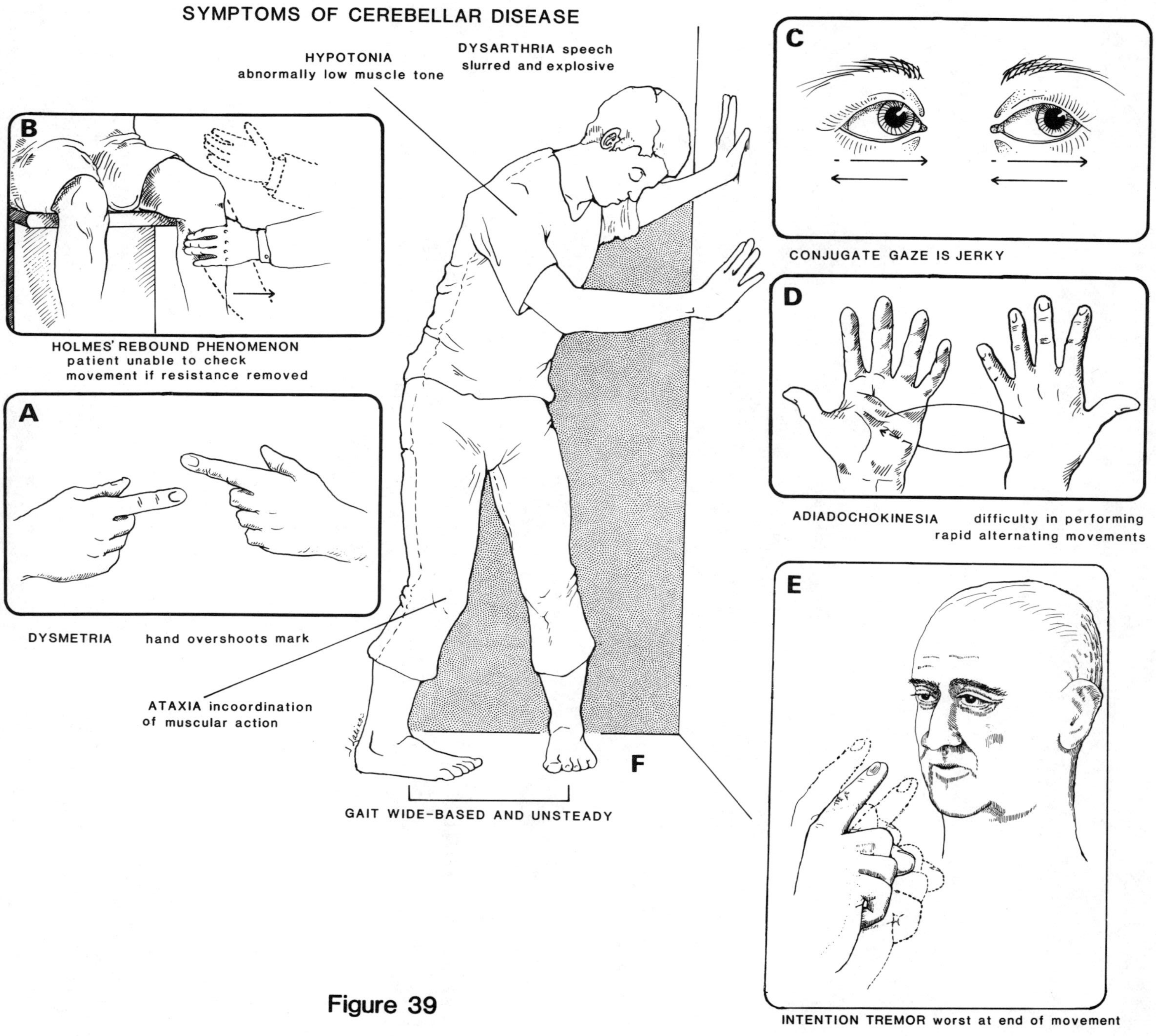

SYMPTOMS OF CEREBELLAR DISEASE

HYPOTONIA abnormally low muscle tone

DYSARTHRIA speech slurred and explosive

B HOLMES' REBOUND PHENOMENON patient unable to check movement if resistance removed

A DYSMETRIA hand overshoots mark

ATAXIA incoordination of muscular action

C CONJUGATE GAZE IS JERKY

D ADIADOCHOKINESIA difficulty in performing rapid alternating movements

E INTENTION TREMOR worst at end of movement

F GAIT WIDE-BASED AND UNSTEADY

Figure 39

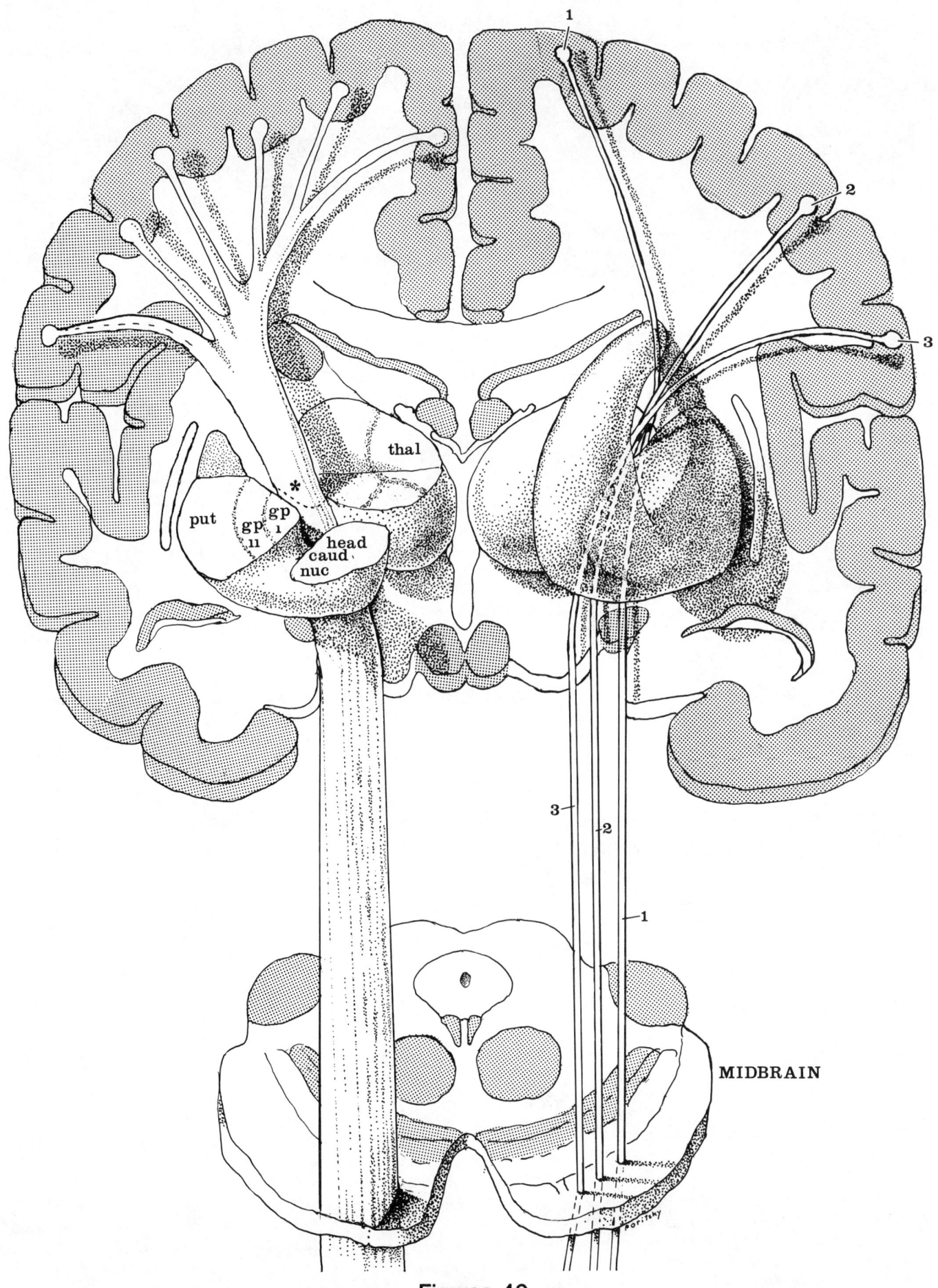

Figures 40

Figure 41

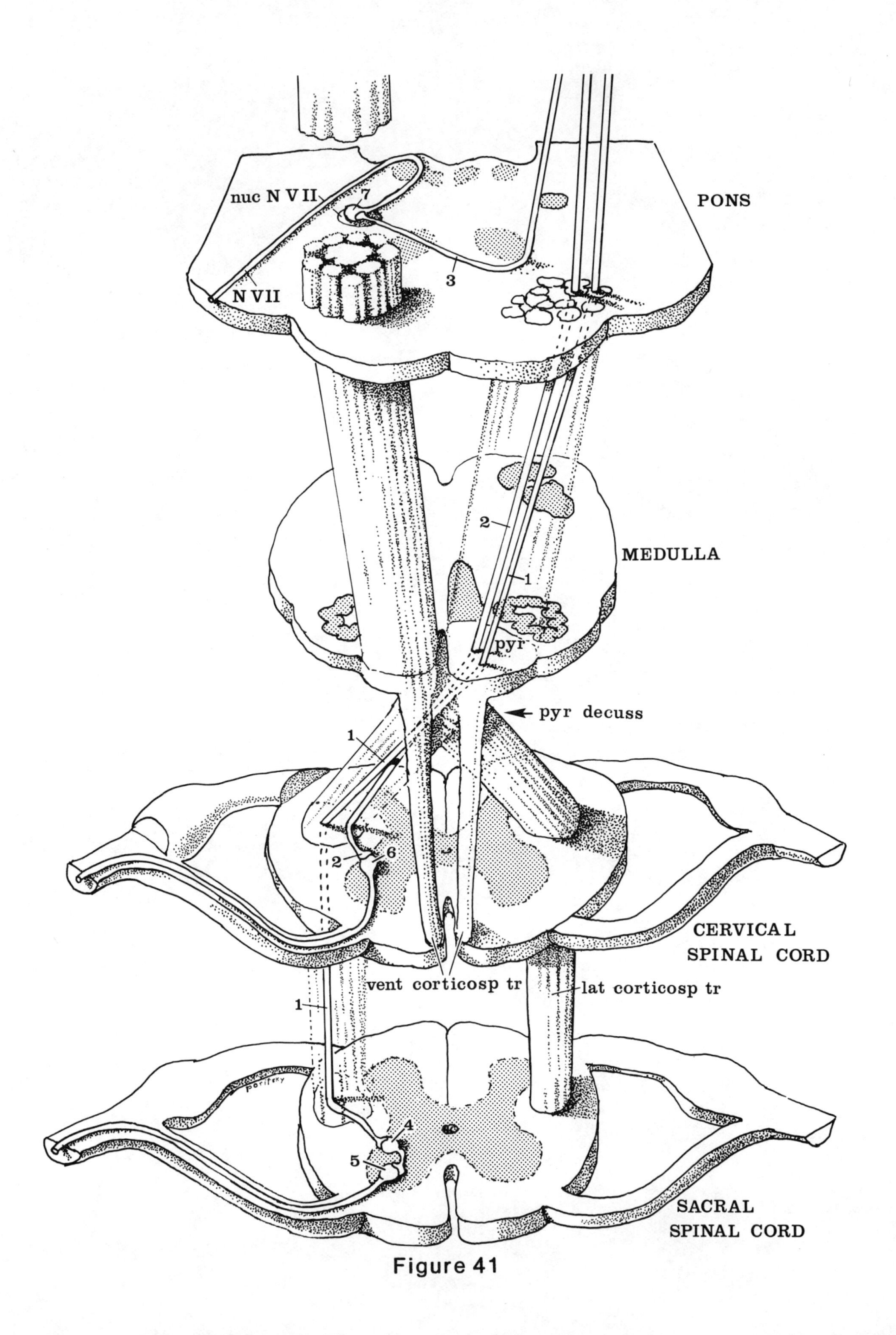

Figure 42

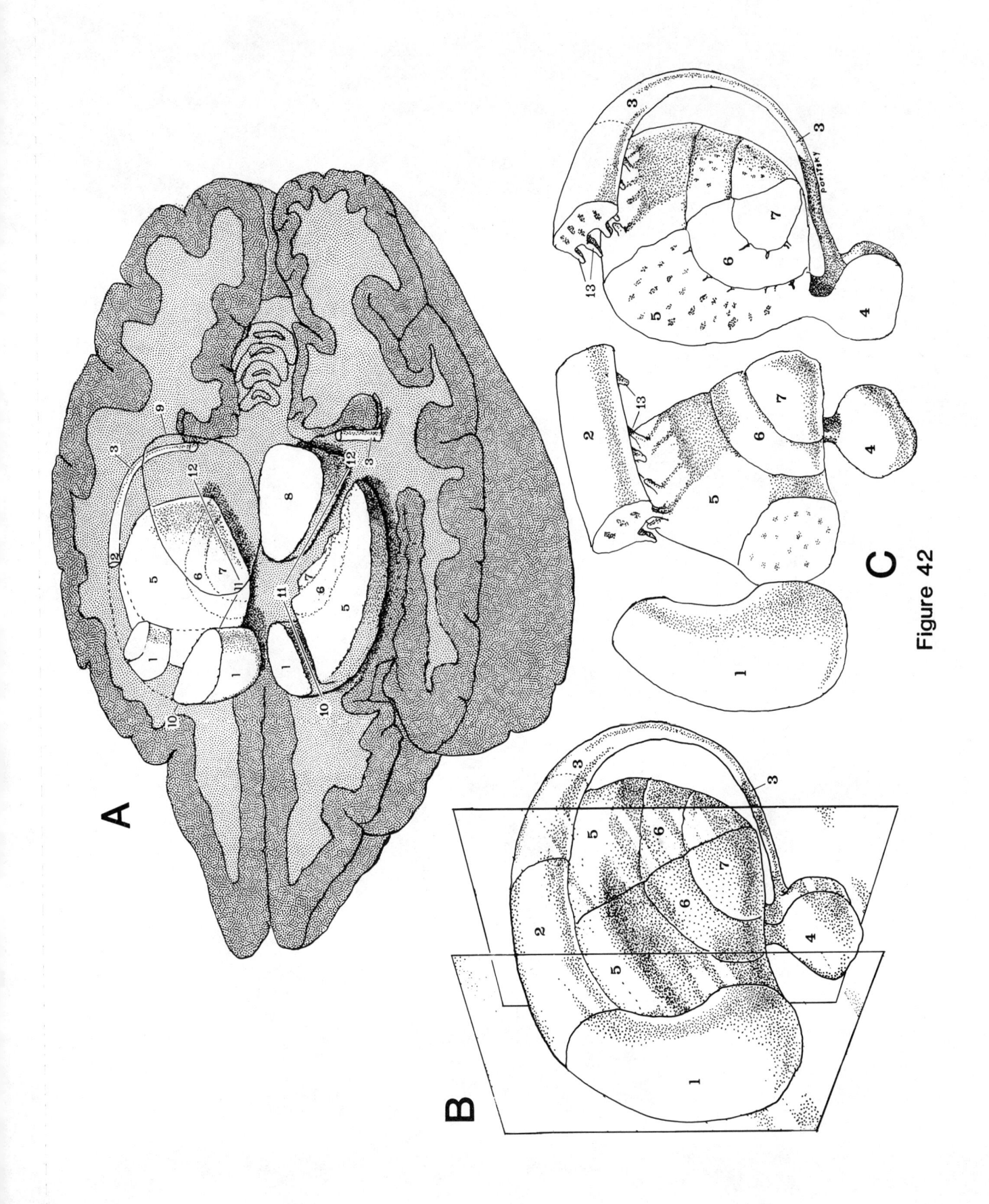

Figure 43

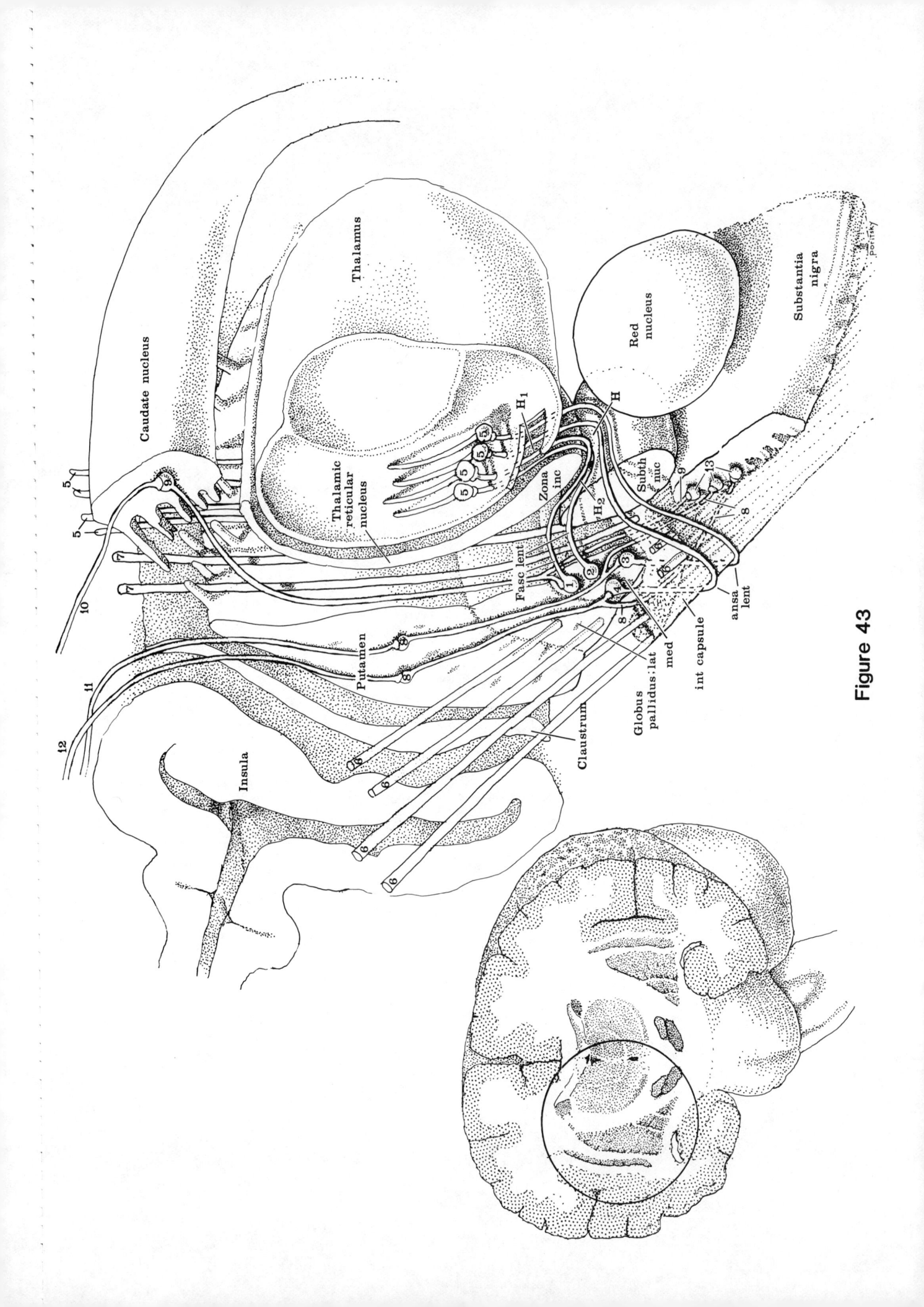

Figure 44

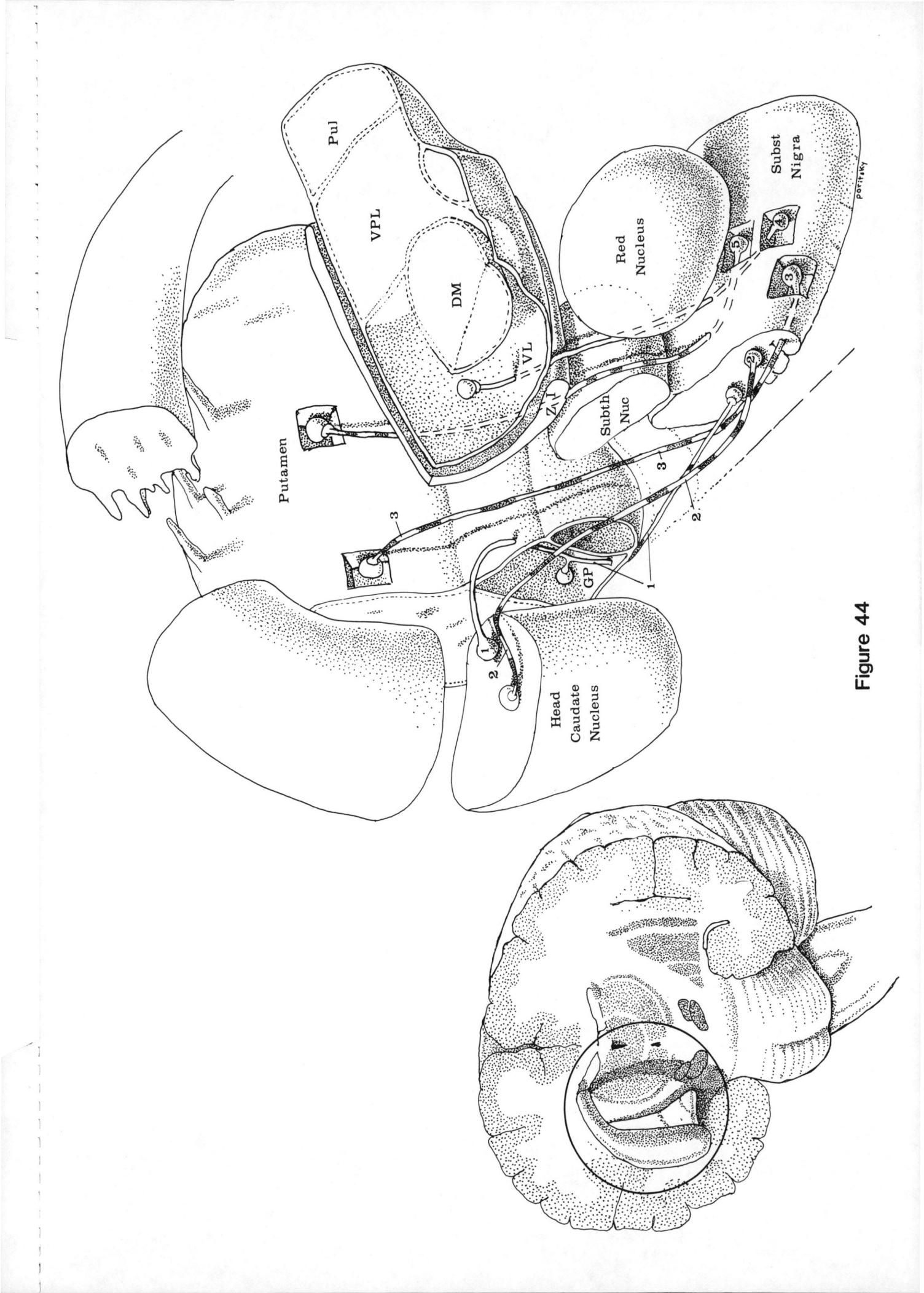

PARKINSON'S DISEASE

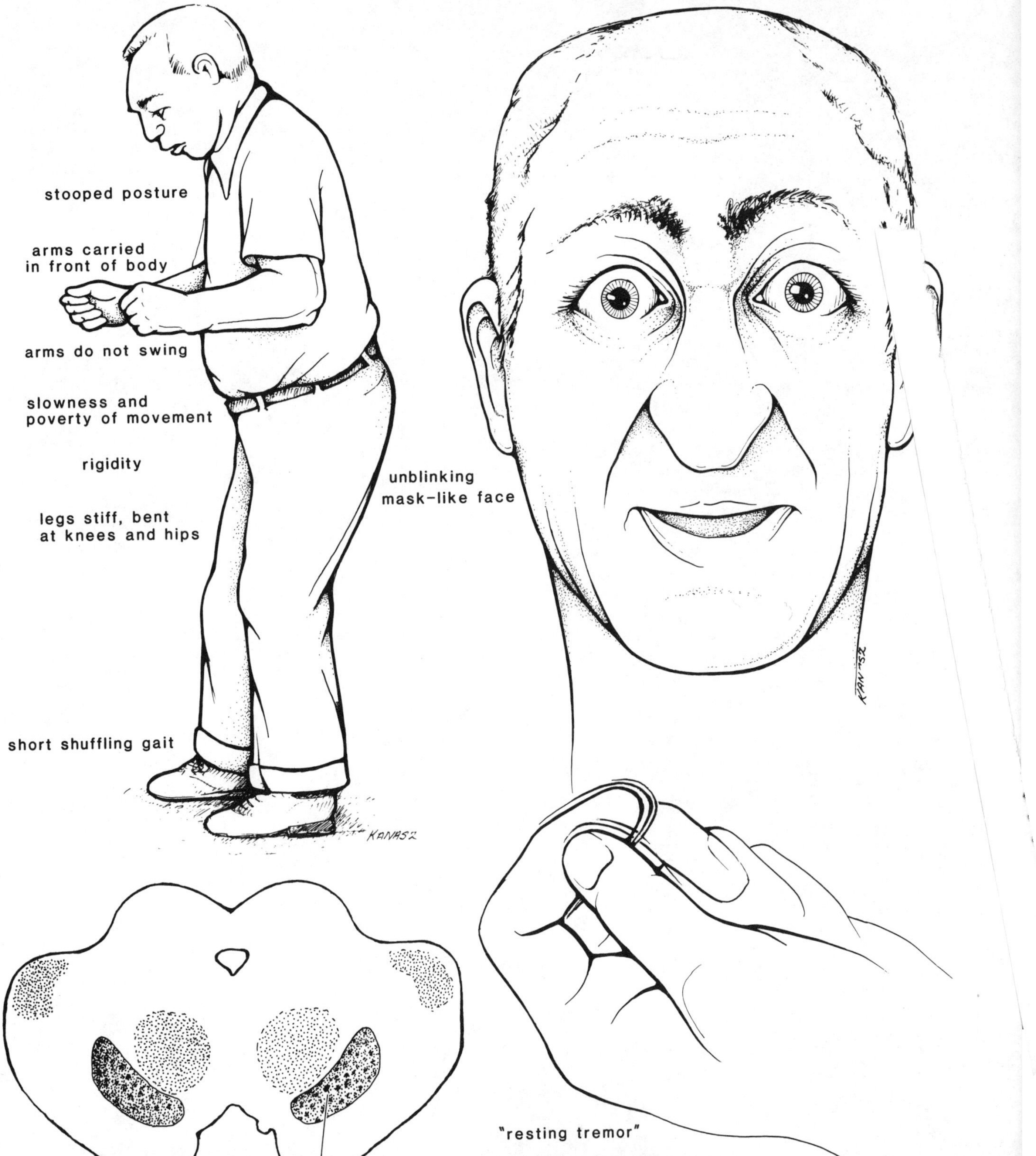

stooped posture

arms carried in front of body

arms do not swing

slowness and poverty of movement

rigidity

legs stiff, bent at knees and hips

unblinking mask-like face

short shuffling gait

degeneration of pigmented cells in substantia nigra

"resting tremor"

often "pill-rolling" tremor in thumb and fingers

Figure 45

CHOREA

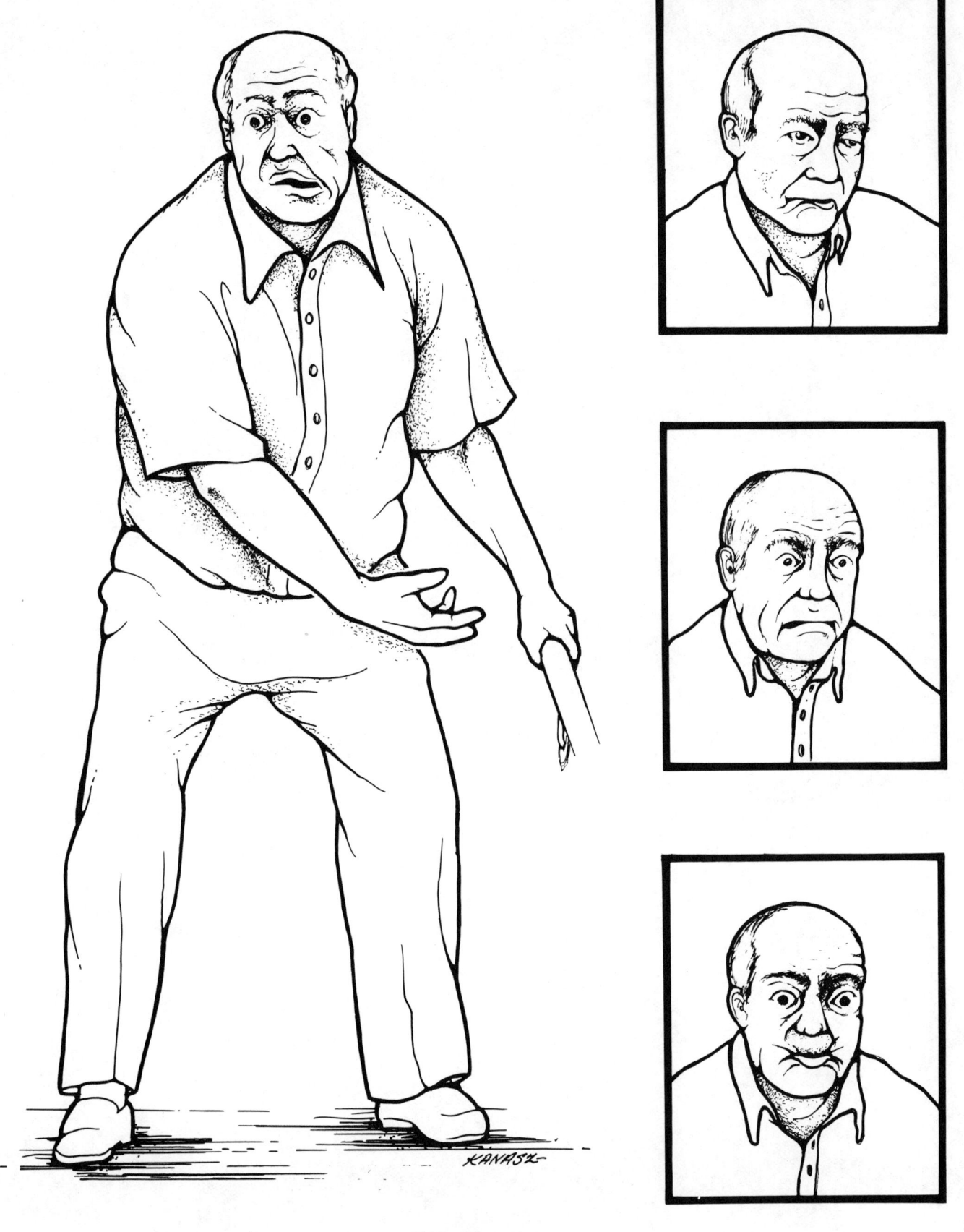

Figure 46

HEMIBALLISM

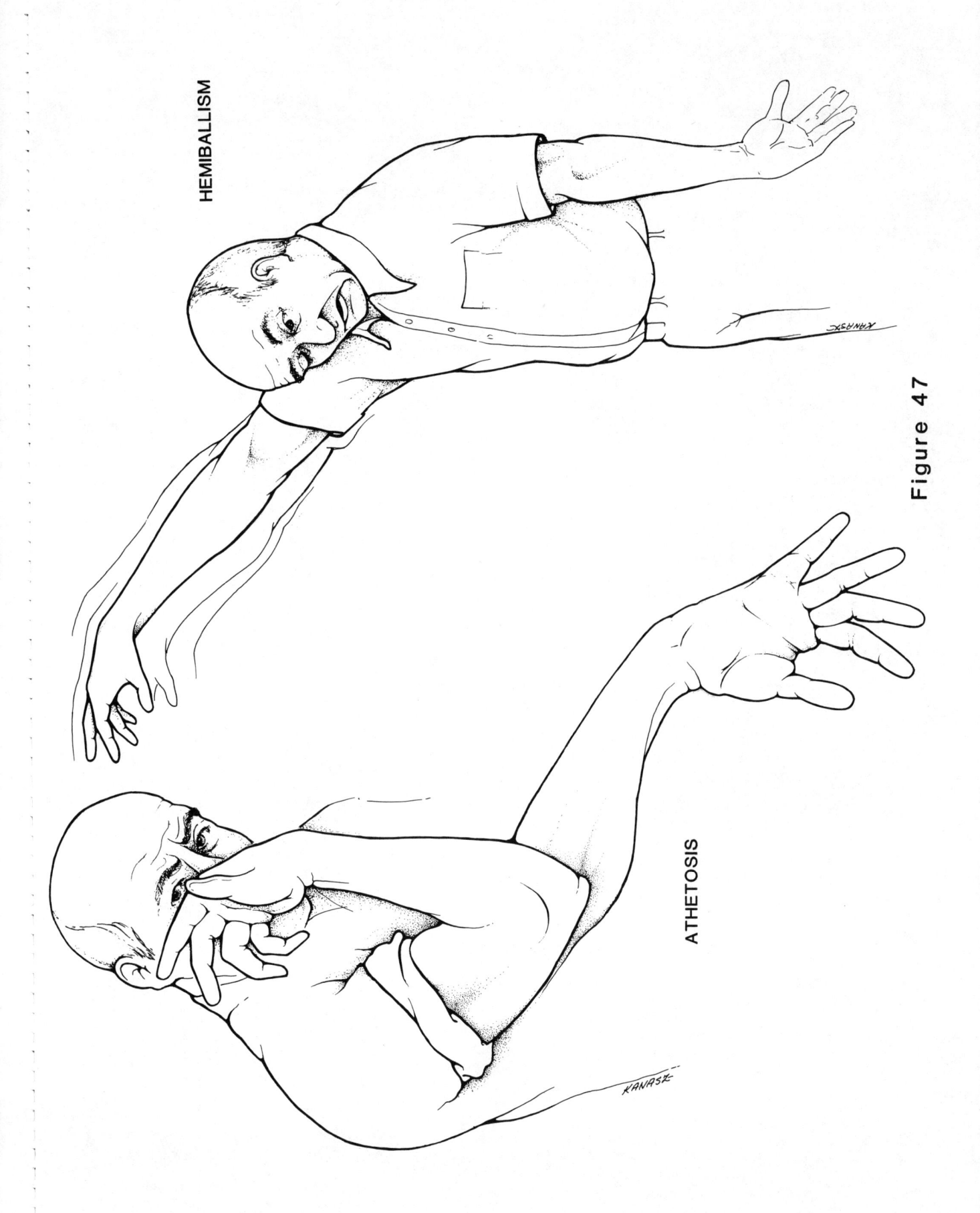

ATHETOSIS

Figure 47

Figure 48

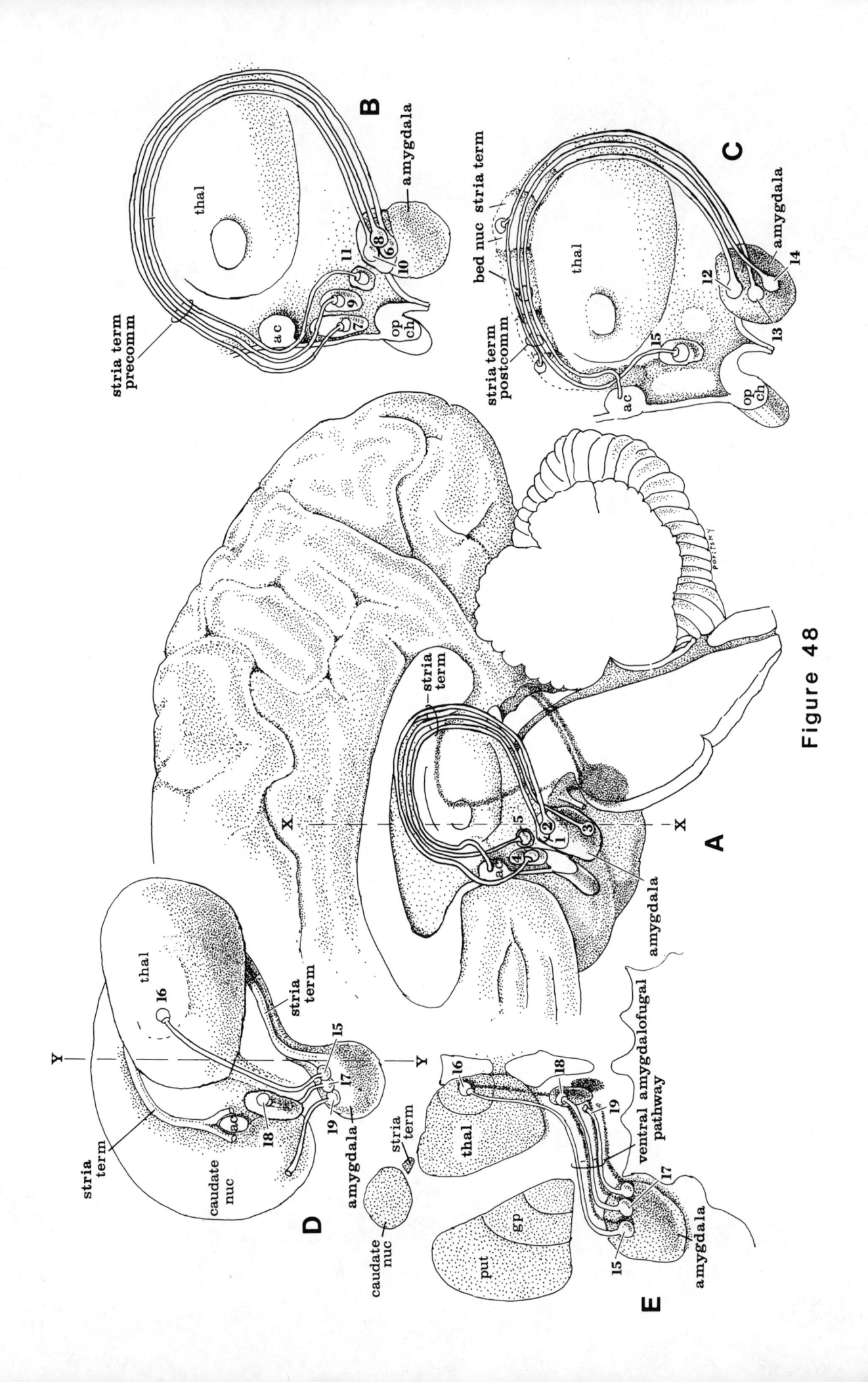

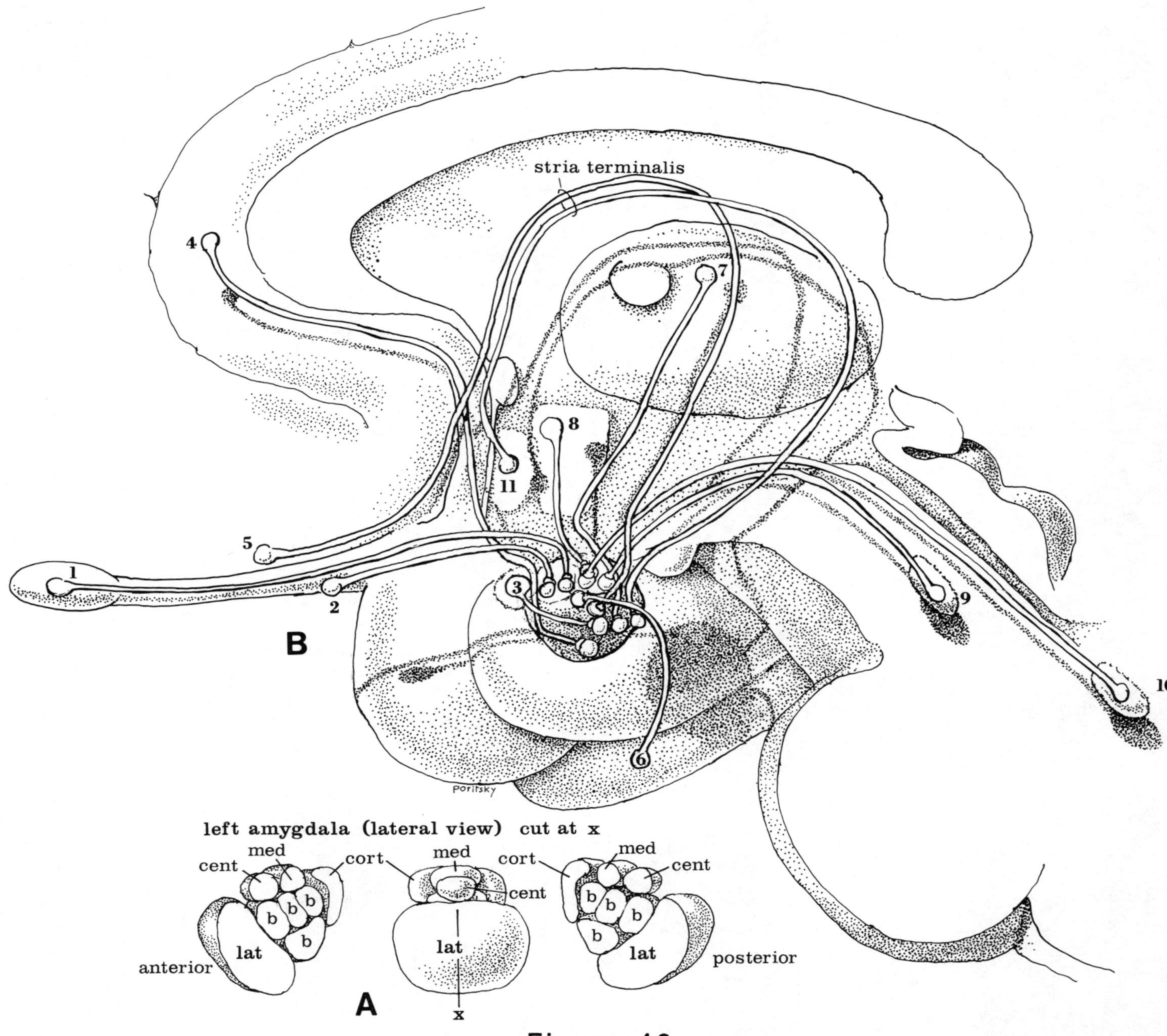

Figure 49

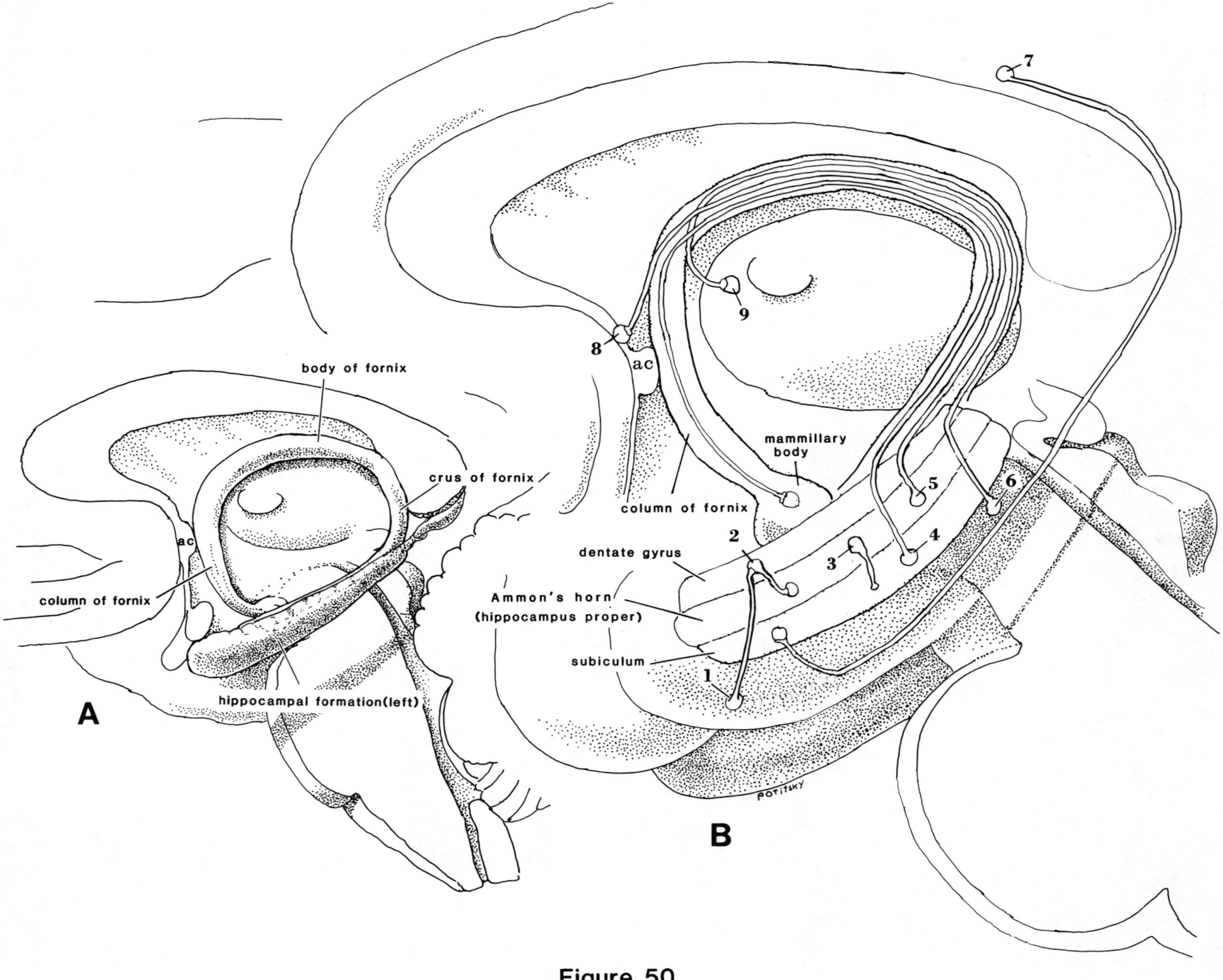
body of fornix
crus of fornix
ac
column of fornix
hippocampal formation(left)
A
7
8
ac
9
mammillary
body
column of fornix
5
6
2
4
3
dentate gyrus
Ammon's horn
(hippocampus proper)
subiculum
1
POTITSKY
B

Figure 50

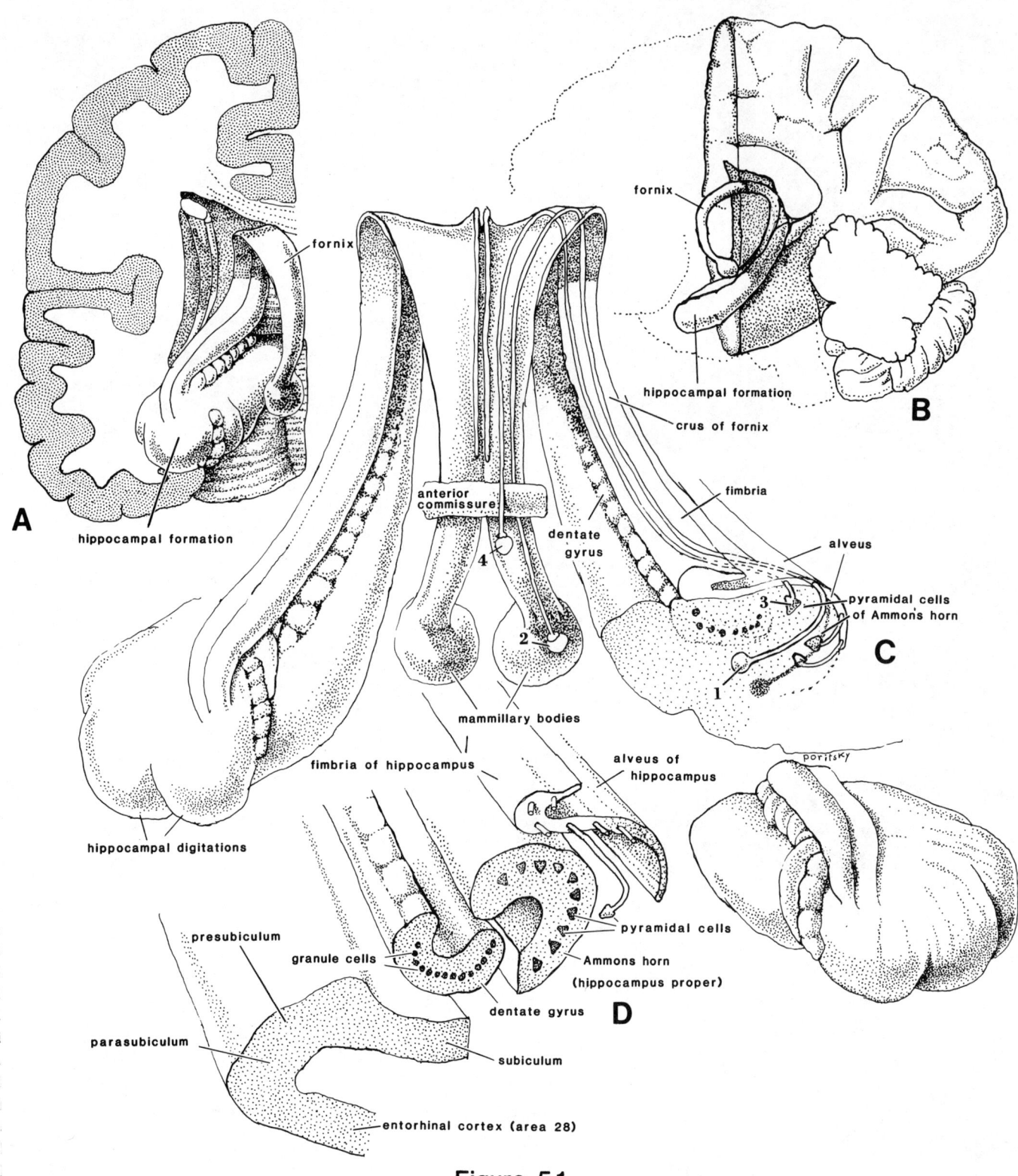

Figure 51

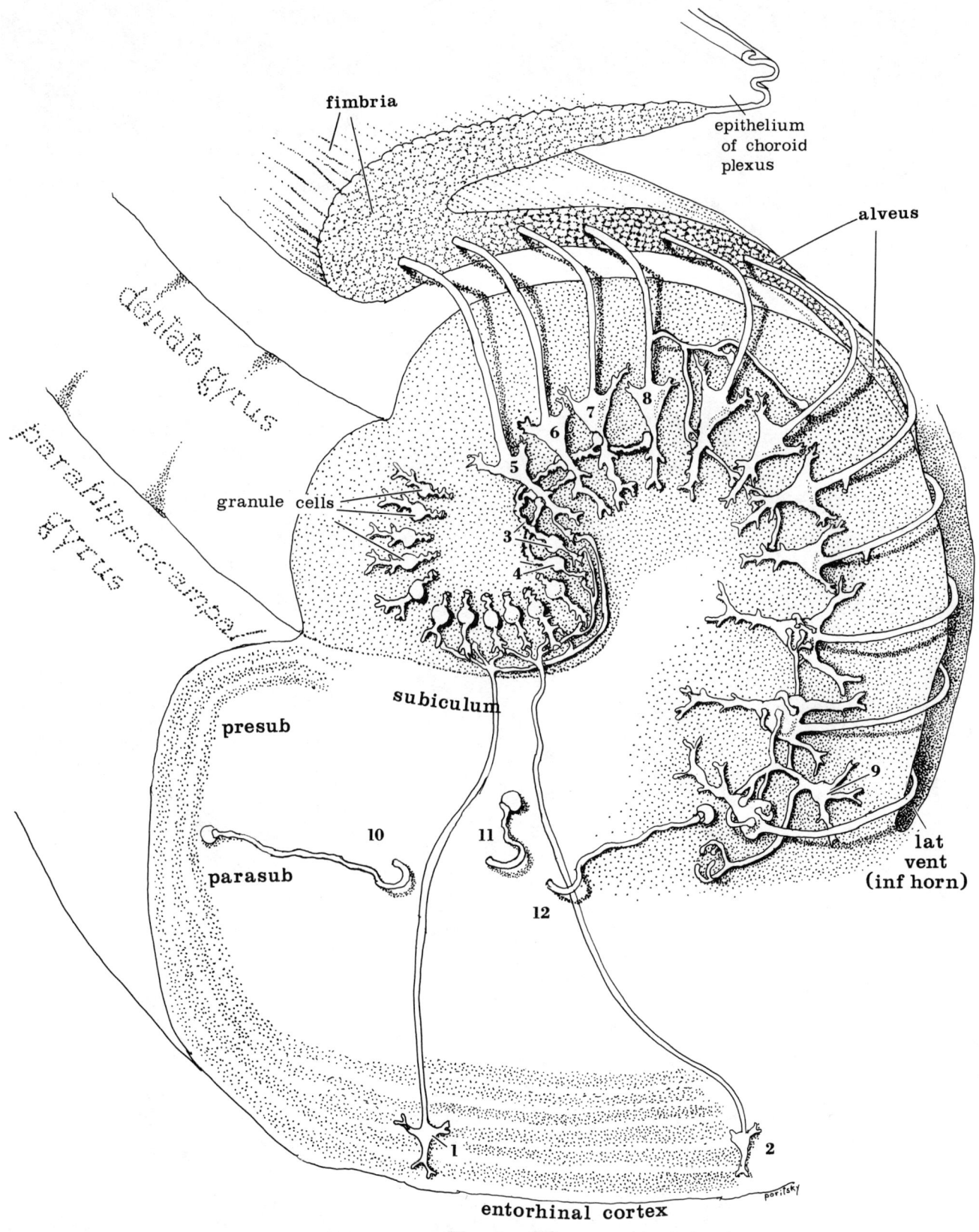

Figure 52

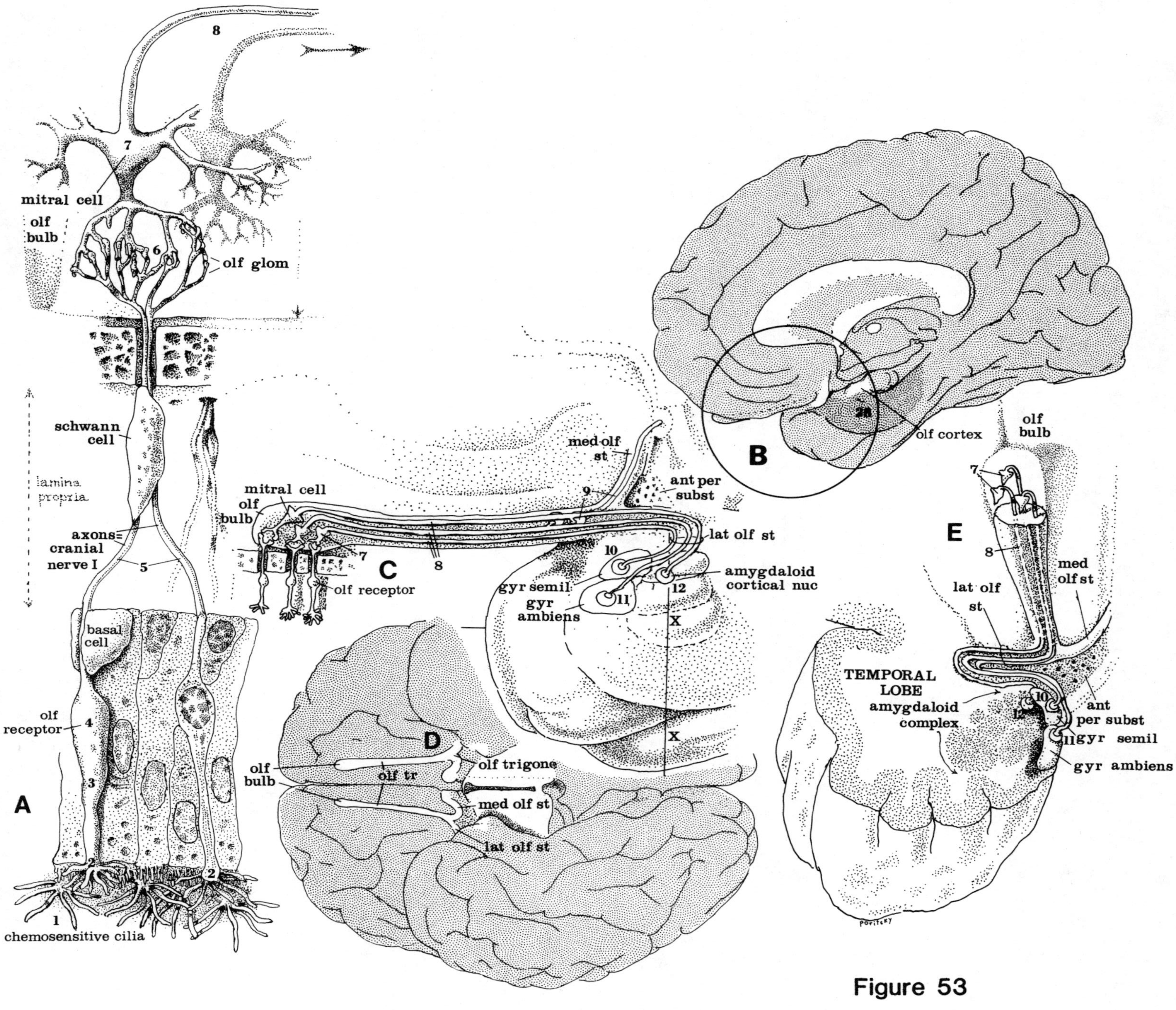
A
mitral cell
olf bulb
olf glom
schwann cell
lamina propria
axons= cranial nerve I
basal cell
olf receptor
chemosensitive cilia
B
olf cortex
C
mitral cell
olf bulb
olf receptor
med olf st
ant per subst
lat olf st
amygdaloid cortical nuc
gyr semil
gyr ambiens
D
olf bulb
olf tr
olf trigone
med olf st
lat olf st
E
olf bulb
med olf st
lat olf st
TEMPORAL LOBE
amygdaloid complex
ant per subst
gyr semil
gyr ambiens

Figure 53

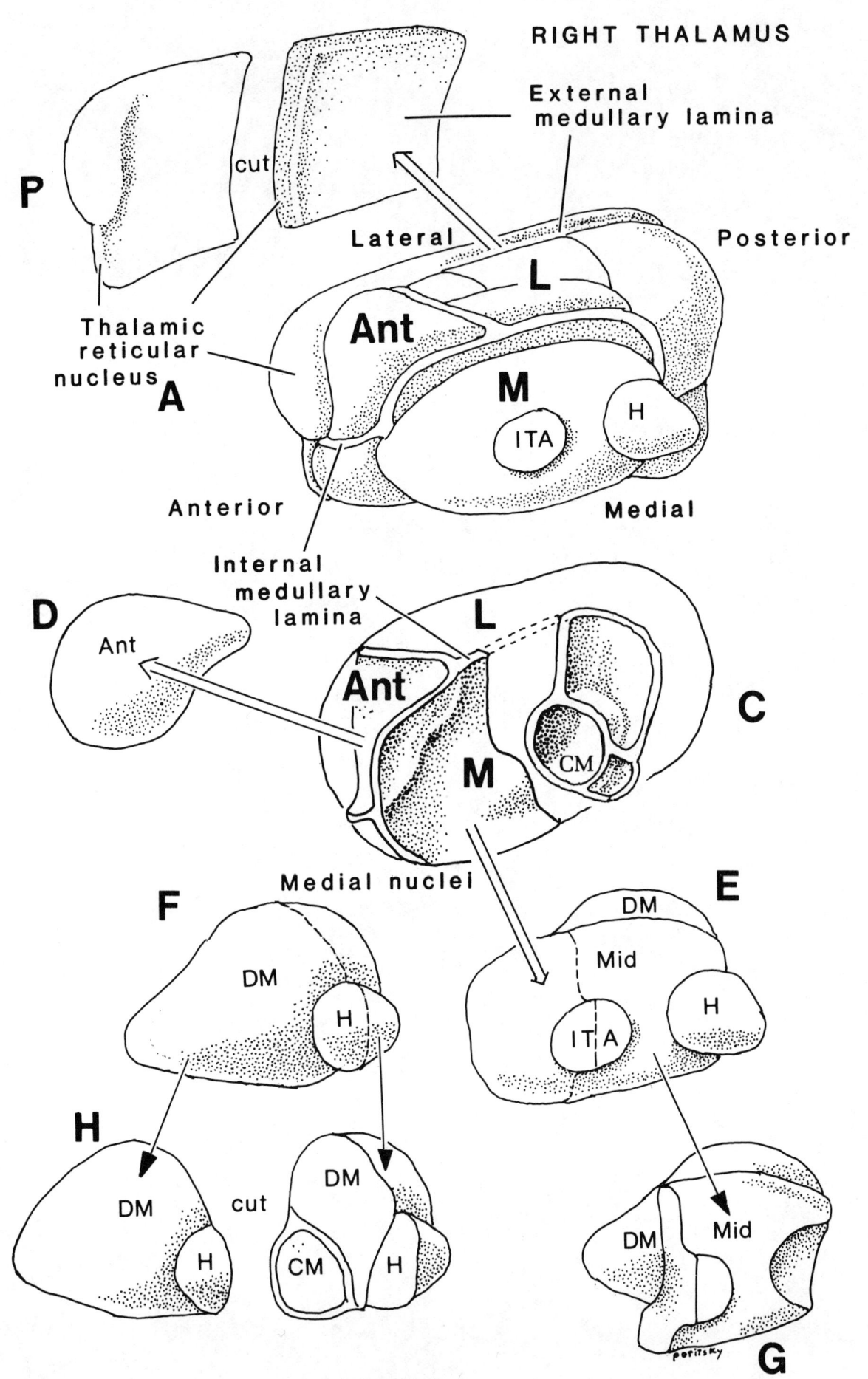

Figure 54

Figure 55

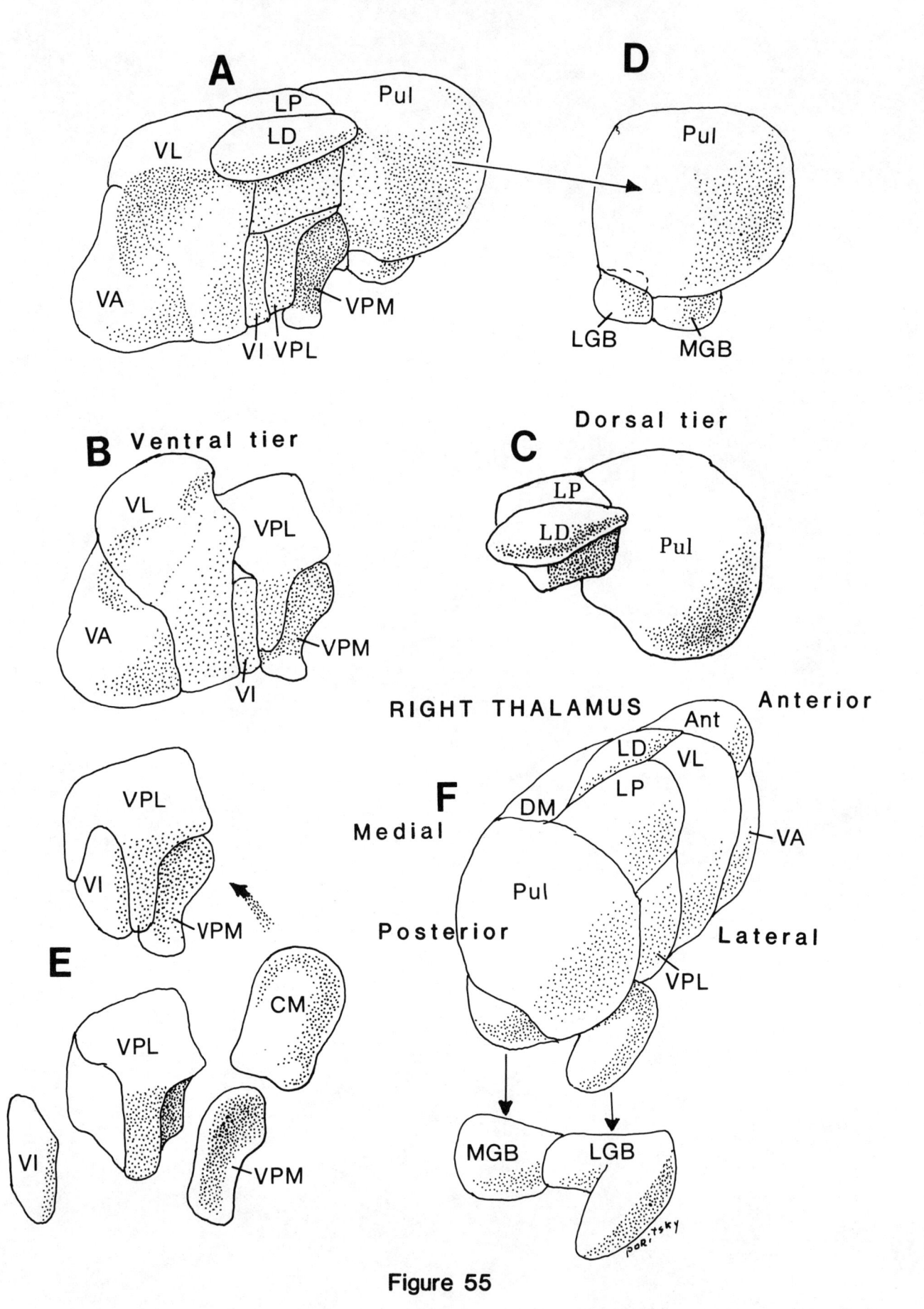

Figure 56

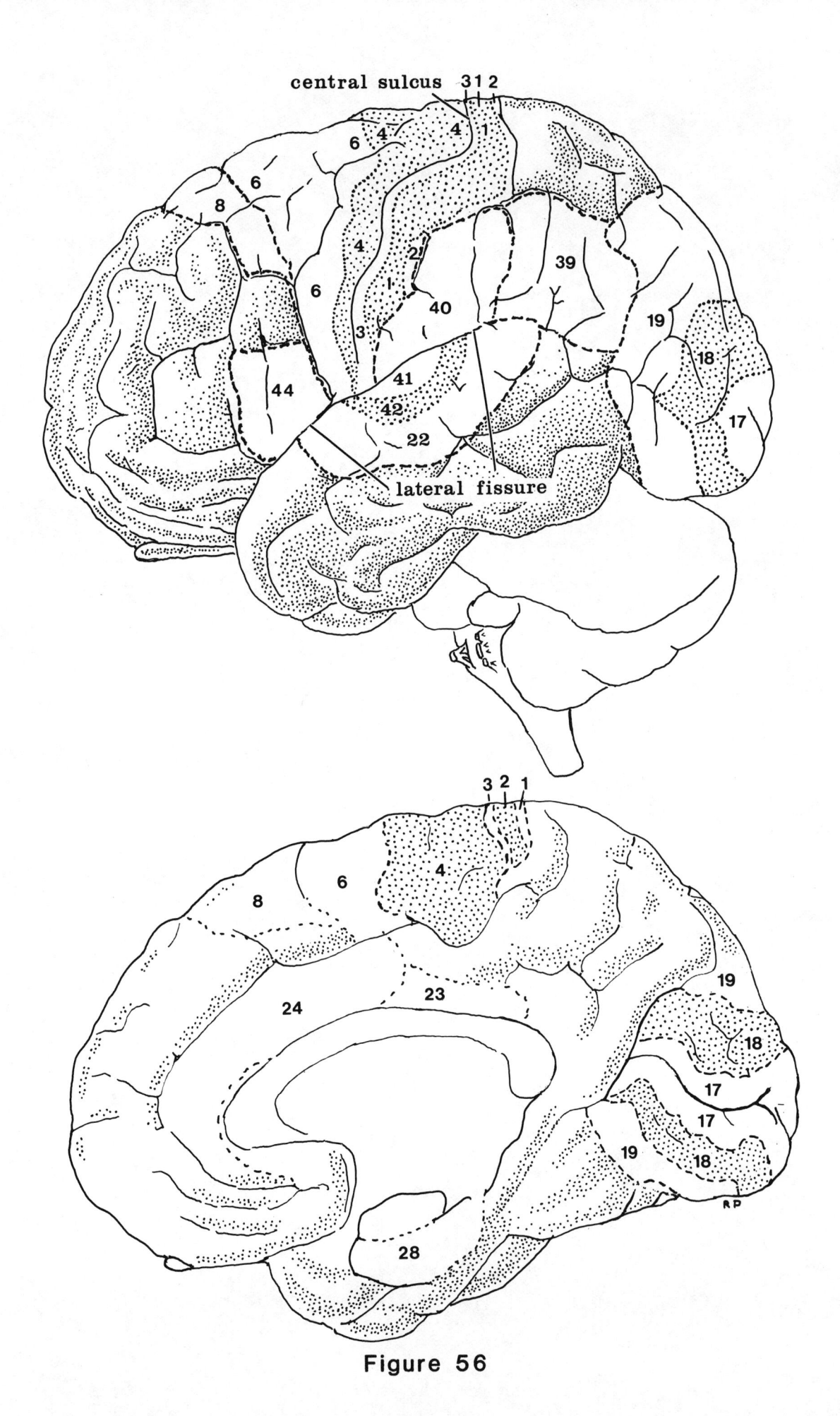

THE STRANGE CASE OF PHINEAS P. GAGE

...while in the act of tamping in an explosive charge of gun powder...

his tamping iron struck fire upon a rock causing the gun powder to explode and shooting the iron bar completely through his head...

...wonderful feats and hairbreadth escapes without any foundation except in his fancy...

his men (with whom he was a great favorite) took him in their arms......

...his mind was radically changed so decidedly that his friends and acquaintances said he was "no longer Gage".......I dressed him, God healed him.

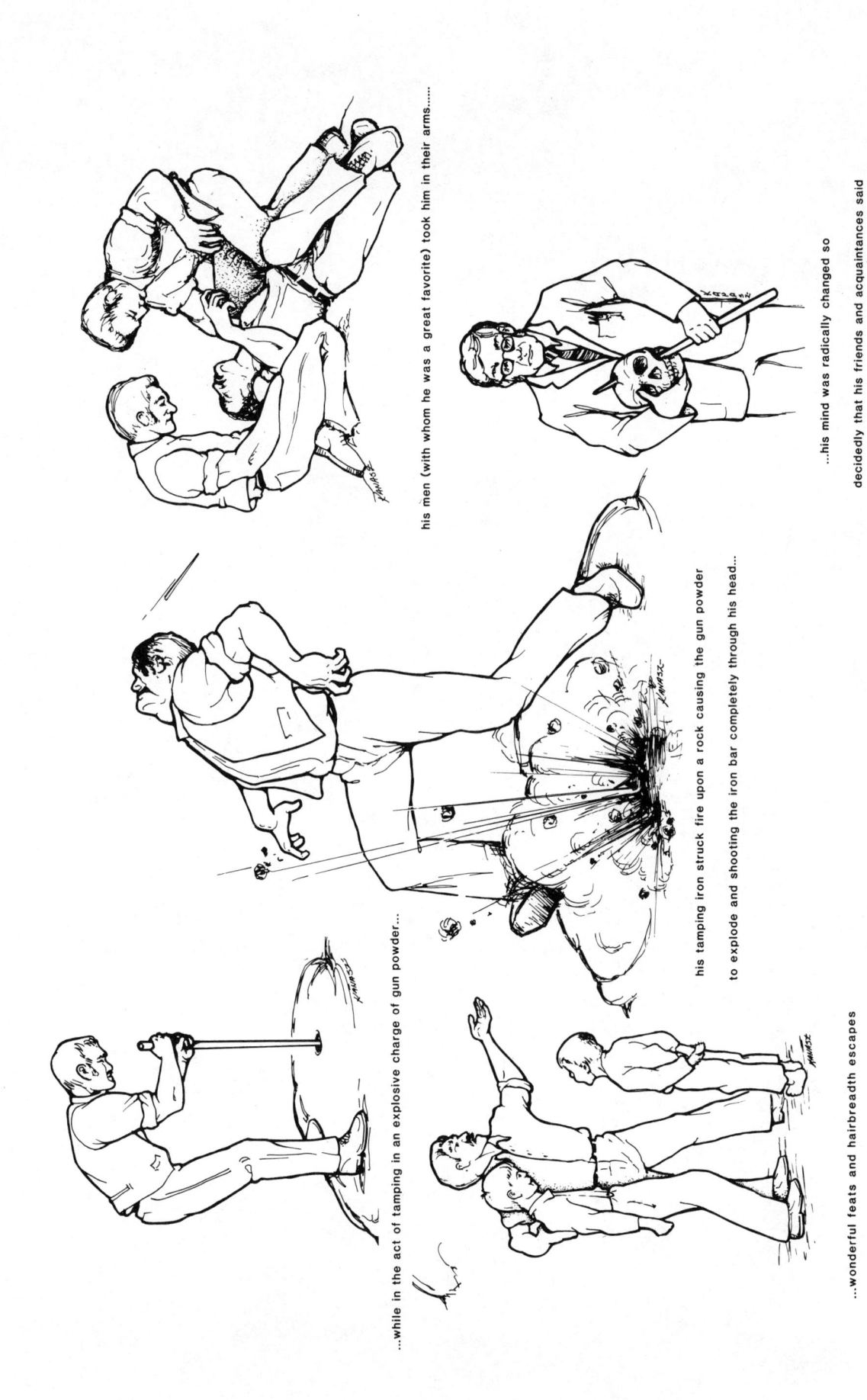

Figure 57

Our first conscious muscular act is nursing at the breast. The neuronal pathways mediating and stimulated by nursing form the beginnings of our awareness of "self" as well as the neuronal substratum upon which all future emotional and mental experience is interpreted and recorded. The tactile and oral sensations that accompany this extremely important act, namely pleasure, warmth, and security are conveyed centrally primarily by the trigeminal nerve, the trigeminal sensory nuclei in the brain stem, the trigeminal tracts, the nucleus VPM in the thalamus and its myriad connections.

Conceivably whether a person is basically happy and content in life, whether he or she is trusting of other human beings, and whether he or she is capable of loving another human being may all depend upon sufficient stimulation, activation and persistence of the these neurons, their connections and their neurotransmitters.

Drawings by Barbara Davison

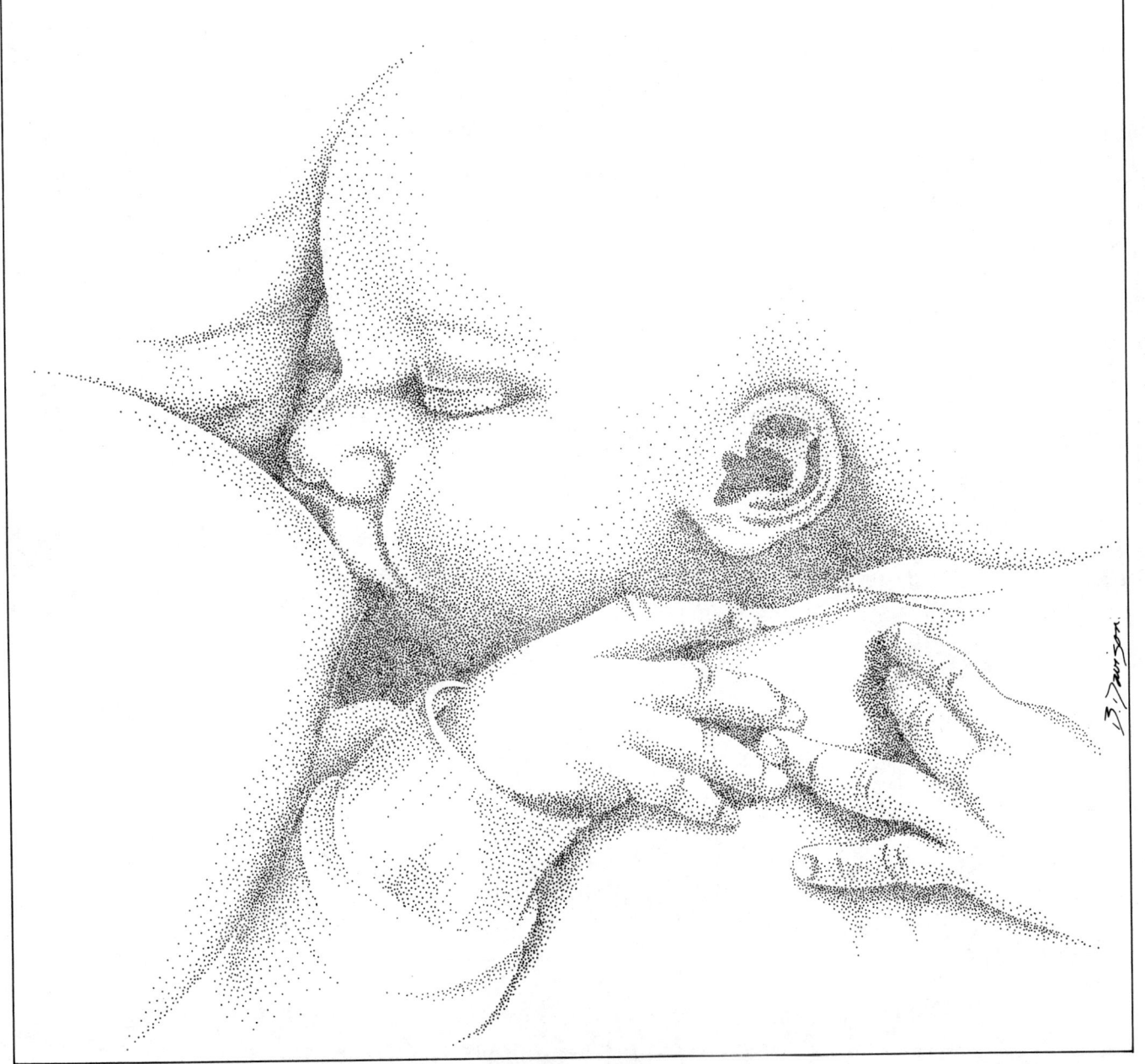

Figure 58